COMPUTATIONAL
BIOLOGY

A HYPERTEXTBOOK

SCOTT T. KELLEY
Department of Biology
San Diego State University
San Diego, California

AND

DENNIS DIDULO
Becton, Dickinson and Company
San Diego, California

COMPUTATIONAL BIOLOGY

A HYPERTEXTBOOK

ASM
PRESS

Washington, DC

Library of Congress Cataloging-in-Publication Data
Names: Kelley, Scott T. (Scott Theodore), author. |
Didulo, Dennis, author.
Title: Computational biology : a hypertextbook / Scott T. Kelley, Department
of Biology, San Diego State University, San Diego, California, and Dennis
Didulo, Becton, Dickinson and Company, San Diego, California.
Description: Washington, DC : ASM Press, [2018] | Includes index.
Identifiers: LCCN 2017051454 (print) | LCCN 2017052307 (ebook) | ISBN
9781683670032 (ebook) | ISBN 9781683670025 (pbk.)
Subjects: LCSH: Computational biology.
Classification: LCC QH324.2 (ebook) | LCC QH324.2 .K45 2018 (print) | DDC
570.285–dc23
LC record available at https://lccn.loc.gov/2017051454

Address editorial correspondence to
ASM Press, 1752 N St., N.W.,
Washington, DC 20036-2904, USA

Send orders to ASM Press, P.O. Box 605, Herndon, VA 20172, USA
Phone: 800-546-2416; 703-661-1593
Fax: 703-661-1501
E-mail: books@asmusa.org
Online: http://www.asmscience.org

To Kina and Aidan, my wonderful and supportive family.

And to my brother Brian, who selflessly donated his kidney, without
which I would not have had the energy
to write this book.

CONTENTS

Preface ix
For the Instructor xi
For the Student xiii
Acknowledgments xiv
About the Authors xv

CHAPTER –1 **Getting Started 1**

CHAPTER 00 **Introduction 5**
Activity 0.1: Biological Databases and Data Storage 20

CHAPTER 01 **BLAST 31**
Activity 1.1: BLAST Algorithm 36

CHAPTER 02 **Protein Analysis 47**
Activity 2.1: Hydrophobicity Plotting 52
Activity 2.2: Protein Secondary Structure Prediction 58

CHAPTER 03 **Sequence Alignment 67**
Activity 3.1: Dynamic Programming 74

CHAPTER 04 **Patterns in the Data 91**
Activity 4.1: Protein Sequence Motifs 94
Activity 4.2: Position-Specific Weight Matrices 102

CHAPTER 05 **RNA Structure Prediction 111**
Activity 5.1: RNA Structure Prediction 118

CHAPTER 06 **Phylogenetics 133**
Activity 6.1: Phylogenetic Analysis 140

CHAPTER 07 **Probability: All Mutations are not Equal (-ly Probable) 157**
Activity 7.1: Generating PAM and BLOSUM Substitution
Matrices 163

CHAPTER 08 **Bioinformatics Programming: A Primer 179**

Index 191

PREFACE

This textbook is a hypertextbook. Half of the textbook material lies between the pages of this book and the other half on the Internet. It seems natural that a hypertextbook, which combines print and online apps for mobile technology, would be a great way to learn the basics of bioinformatics, which uses informatics (computational) theory to study biological data.

This book was born out of a mix of necessity and inspiration.[1] The necessity came from the dearth of bioinformatics instructional materials appropriate for my combination of biology students, with little or no computer background,[2] and computer science students, who were interested in the field but had little understanding of biology. The need became acute when I learned that my favorite bioinformatics lab manual, *Bioinformatics for Dummies* (BFD[3]), would no longer be updated. BFD was a great lab manual for learning how to perform basic bioinformatics data analysis. This book did not explain the principles behind the algorithms, but I could cover those during lectures. BFD was clear and fun to read and provided practical skills for biologists and others looking to analyze data. Unfortunately, the most recent version was printed in 2007!

I kept using the old edition of BFD for some time, but eventually the tutorials became obsolete and the students took longer and longer to complete the exercises. In fact, several passages of BFD were obsolete a few months after the book was printed. Bioinformatics websites are constantly changing, including their designs and the URL links, and sometimes the pages themselves disappear altogether. Since I began writing this book, two of the websites I teach in the book and online materials changed significantly, and one disappeared altogether.

This led to my original inspiration for the hyper- part of this hypertextbook. What if I made my own bioinformatics tutorials and sample test data for commonly used analysis tools online in easily updated files? That way, when a link changed or the programmers moved a radio button around, I could easily alter the tutorial to reflect these changes in real time. Students would not have to wait for a new version of a book to have an accurate tutorial.

The next inspiration arose from my use of paper-based puzzles and problems to teach the bioinformatics algorithms. The problems I taught in class, combined with the anticipatory conceptual exercises and lecture material, were very successful for teaching how the methods worked. Unfortunately, paper-based problems also had significant drawbacks: the students were given only one practice problem per algorithm, and they received very little feedback as a result. Typically, I would (1) teach the method, (2) do an example with the students in class, (3) assign it for homework, (4) get it back a week later, and (5) return it with feedback a week after that. And that was it.

Fortunately, I realized that the structure of the algorithm puzzles I taught would be perfect for touchscreen devices and laptops. Most of them involved either sliding letters around or filling in boxes with numbers, both easily done with a finger or a mouse. With the collaboration of my bioinformatics website designer and coauthor Dennis Didulo, I created interactive learning tools that provide limitless practice and instant feedback for students. When we combined the bioinformatics software tutorials and test data into one site, we had a comprehensive learning paradigm for introductory bioinformatics. (See "For the Student" section below for an outline of the website features.)

In my class, I noticed an immediate increase in algorithm comprehension and problem-solving ability. Students gained much more practice, received more feedback, and performed much better on tests. Because new problems were easily randomly generated, each student had their own personal data set. Best yet, I could now quickly generate new exam questions and answers with the click of a button!

And what did my students think? These quotes speak for themselves.

"It is a wonderful learning tool. The online programs made learning the algorithms almost easy."—Ruby, undergraduate student

"I didn't want to tell you how much I liked the website because I didn't want your ego to get too big."—Emily, undergraduate student

"It was much better than that bioinformatics cat video."—Pedro, graduate student

"You can learn bioinformatics while waiting in line at the DMV or sitting on your couch eating cheese puffs!"—Anonymous

Notes

1. Much like the invention of the salad spinner.
2. Many biology students tell me flatly that they are "bad with computers" or even state that "computers hate [them]." For the record, computers really don't care about you at all. Which is why we should never give them weapons (see the film "Terminator").
3. BFD, the budding bioinformatician's BFF.

FOR THE INSTRUCTOR

This hypertextbook can be used in a number of ways: in a lecture or online course, using the book as an outline for a course, or using just the sections of interest. It is important to note that, being a hypertextbook, the web components are not supplemental, but instead are crucial for being able to understand the content presented in the physical book. In my classes, I use the interactives inside of class, and the students also use them outside of class to help them solve algorithm problems or prepare for exams. Generally, I use the following approach:

1. Teach the biological relevance and background of the method.

2. Have students solve the conceptual (anticipatory) exercise in class, sharing answers with one another.

3. Lecture on the basics of the algorithm.

4. Have students bring out their mobile devices (laptops, smartphones, and tablets) and solve the interactive problems.

5. Have students share their answers with neighbors in class and with the instructor.

Then, to make sure the students practice at home, I assign the paper-based practice problems. Finally, in the computer lab, or for homework, I assign the lab exercises with the software based on the algorithms.

So far, the approach has been a great success in my classes. The online tools increase comprehension and improve exam results, and the easily updated tutorials for bioinformatics analysis software and biological databases reduce a lot of student frustration. I hope it proves as successful in your class as it has in mine.

FOR THE STUDENT

This textbook is really a hypertextbook, meaning that much of the most exciting learning happens online. Close to half of the book materials are online, and in each chapter you will be directed to the online tools associated with the text. The idea is to leverage the uniquely powerful aspects of the Internet to help you learn about bioinformatics. The puzzle-like nature of bioinformatics algorithms makes them especially suited to interactivity and "gamification" (making difficult things into games with points and scores). The interactive nature of mobile devices and their connection to online bioinformatics software make them useful learning tools for understanding the theory behind bioinformatics methods (the algorithms) and for gaining practical experience with their implementation (software analysis and databases). In order to enhance learning of the principles behind bioinformatics algorithms and make them more engaging, the online resources have been designed to

- **Be interactive**, with touchscreen puzzle-like problem sets that provide instant feedback

- **Be multiplatform**, usable on computers, tablets, and smartphones

- **Be highly practical**, with direct links to data analysis websites and including test data sets and step-through tutorials

- **Be easily updated**, because bioinformatics websites change constantly and tutorials often need adjustment

- **Allow plenty of practice** through instant "random" problem generation and quizzes

ACKNOWLEDGMENTS

wish to thank the leadership of the California State University Program in Education, Research, and Biotechnology (CSUPERB) and the grant reviewers who approved my proposal on mobile app education technology that provided the seed money for developing the interactive technology and web resources. I thank Greg Payne at ASM Press for listening to my ideas and taking them seriously and for his support during the writing and publishing process, and I thank my colleague at SDSU, David Lipson, for telling Greg about my project. I thank the hundreds of bioinformatics students who took my course at SDSU, who helped me refine my algorithm teaching methods from their sub-alpha development pencil-and-paper stages all the way through to the interactive app stage. You are the reason I do all this in the first place. I give special thanks to my spouse Kina Thackray for her advice during the long process of developing the bioinformatics learning algorithms, for encouraging me to submit grant proposals, and for her very helpful comments on multiple drafts of the book. Finally, I need to thank my former biometry professor Dr. Michael Grant, who taught me statistics and introduced me to programming (SAS) and Dr. Gary Stormo, who graciously allowed me to pursue bioinformatics as a postdoctoral researcher in his lab.

ABOUT THE AUTHORS

Scott T. Kelley is a Professor of Biology at San Diego State University. He has a Ph.D. from the University of Colorado and a B.A. from Cornell University. His lab at San Diego State University combines phylogenetic methods and culture-independent molecular tools to study environmental microbiology. Dr. Kelley has published extensively on the human microbiome, the built environment, and many natural environments. He has published many papers on bioinformatics, and has helped develop some widely-used tools for analyzing next-generation sequence data sets for  microbial communities. He has received research grants from the National Institutes of Health, the National Science Foundation, the Alexander von Humboldt Foundation, and the Alfred P. Sloan Foundation, among others. He has served on the scientific advisory board of the Clorox Company, and his work has been featured by *The New York Times,* NPR, CBC (Canada), *Time Magazine,* and *Der Spiegel,* among numerous others. He is a massive fan of the FC St. Pauli and Everton FC football clubs; loves punk rock, jazz, and classical music; speaks German for fun; and makes a mean apple pie. You can follow Scott on twitter@kelleybioinfo.

Dennis Didulo has been a Data Analytics/Software Engineer at CareFusion since 2014 and a Software Test Engineer at Becton, Dickinson and Company since 2016 and also teaches online database and programming courses for the University of Maryland University College. He received his master's degree in information technology at De La Salle University and his master's degree in bioinformatics at San Diego State University. Dennis has professional development expertise in more than a dozen computer languages,  as well as expertise in database management, algorithm design, and systems engineering. Dennis is a proud father of five grown children, whom he surprised by flying back unannounced to the Philippines for a visit.

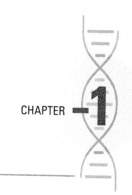

GETTING STARTED

Using the Website

Direct your browser on your phone, computer, or tablet to the following website:
http://www.kelleybioinfo.org

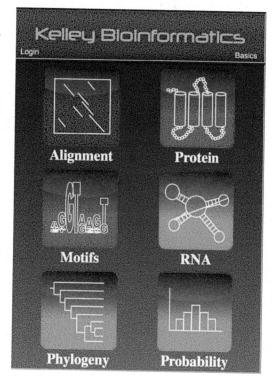

There you will see the homepage, as shown at right.

Touching or clicking an icon (e.g., "Alignment") will take you to a new page that has tools related to the icon topic. The Alignment, Motifs, and Phylogeny buttons teach algorithms and tools for many types of sequence analysis with DNA, RNA, and proteins. The Protein and RNA buttons focus on algorithms for predicting structural features of the functional macromolecules, while the Probability button teaches how to generate substitution matrices.

Example: The Alignment Page

Clicking or touching the Alignment button will take you to the following page, which begins with the BLAST algorithm interactive tool. All the pages use this basic design.

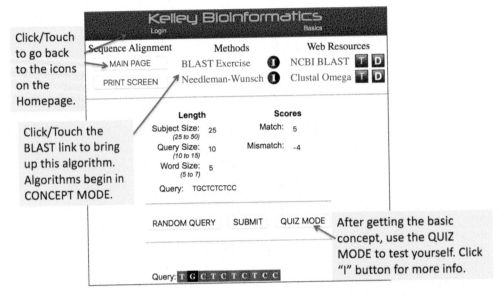

Click/Touch to go back to the icons on the Homepage.

Click/Touch the BLAST link to bring up this algorithm. Algorithms begin in CONCEPT MODE.

After getting the basic concept, use the QUIZ MODE to test yourself. Click "I" button for more info.

General features

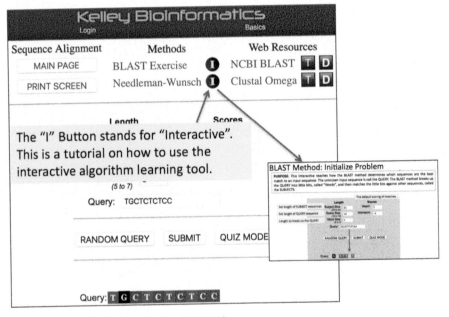

The "I" Button stands for "Interactive". This is a tutorial on how to use the interactive algorithm learning tool.

Information on the interactive learning tool

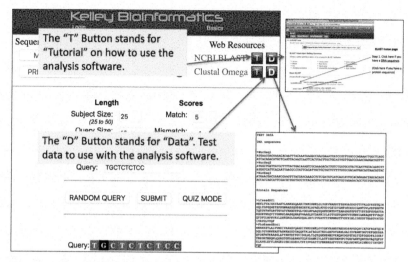

Tutorials and test data for online bioinformatics software

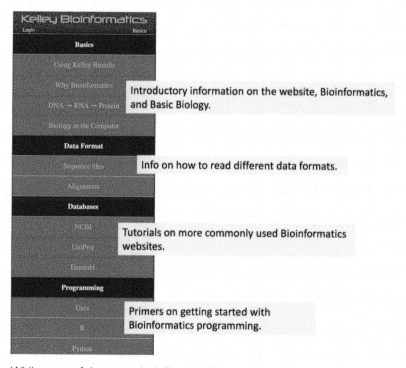

While most of the pages look like the Alignment page, the Basics page is organized differently and mostly contains information and tutorials.

How To Use This Book

I will assume you are familiar with how to read/use a book, but remember that the physical book is meant to be used in conjunction with the online component. Throughout the text you will be directed to online modules via URLs and QR codes. The online material is not supplemental, but is a critical portion of this hypertextbook.

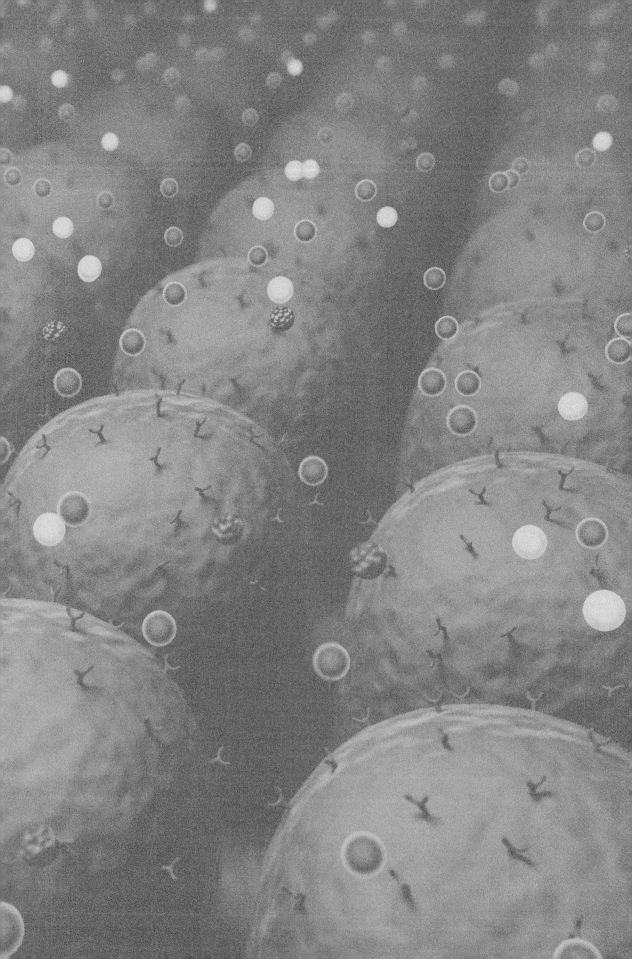

INTRODUCTION

The word bioinformatics refers to the computational analysis of complex biological data. The "bio-" prefix indicates biology, of course, while "informatics" is the science of data processing, storage, and retrieval (a.k.a. information science) that first developed in the 1960s. Bioinformatics itself dates back to the early 1970s, when computers were first used to analyze molecular sequences. While our knowledge of biological processes, the amount of molecular data, and the speed and throughput of computation have all expanded dramatically, the field of bioinformatics still primarily focuses on the analysis of three critical biological molecules: DNA, RNA, and protein.

These molecules are critical to the cellular processes of all living organisms, and the analysis of the composition and patterns of these molecules should in theory reveal all the secrets to life. (Or, as Dr. Frankenstein would say, "It's alive! Bwahaha!") In fact, because DNA encodes the information for all the RNA and protein in every cell, analysis of DNA sequence patterns comprises the majority of bioinformatics. RNA and protein sequences are also analyzed using specific bioinformatics algorithms, but the sequences of these molecules are often computationally determined from the DNA sequence in one way or another (see below).

The purpose of this chapter is to explain the general properties of these biological molecules and how they are represented and stored in the computer. It is critical to understand the connection between the data you observe in computer files and the biological molecules. Otherwise the data analysis and databases that store these data will make little sense. We also briefly discuss what is known as the central dogma of molecular biology, how the DNA inside cells is "read" by the cellular machinery, and the general structure of the gene. The introductions to each chapter provide additional background information about the structure and function of DNA, RNA, and proteins and how bioinformatics can be used to analyze different aspects of these molecules.

Why Bioinformatics?

When nonscientists ask me what I do for a living, I tell them I'm a computational biologist. This oxymoron usually elicits a confused expression ("You study the biology of computers? Say what?"). I quickly follow this by asking them if they have heard of DNA and the human genome, which most people have by now. Then I tell them that the DNA that makes up the human genome is really just 3 BILLION LETTERS in a computer. Here is a little snippet of DNA information from the human genome:

```
AGAAAATCACCCTTCCCAGGGGGAAGGTGCTGGGCAGTGGCACTGCCTCTTGGGGGAAGAGGTTGGGCAG
GGGCTGACGGGCAATGGCAGATGACAGCATCCAAACTTCCACACACAGAGTCTGTTCCTTCCTCTTCCCC
GTGCCATCCCAACTCCCTTCTGCCTTGTCATCTACGTCATGGGAAGCAGGTGACATATCTGGCAAGTTAT
TTTGGGGGCCTGGCTTCTCCCAGGTGAAGAGGGAGCAGCAGCTGGAGGGGCAGAAAGAGGGGACAGGGAG
GGGCTGGAGGGCACAGCTGAAGACAGCCTGGGAGGTGACTGTCATCCCCTCCAGTCTCTGCACACTCCCG
GCTGCAGCAGAGCAGGAGGAGAGAGCACGGCCTGGAATGCTAATTTGCCAGGAGCTCACCTGCCTGCGTC
ACTGGGCACAGACGCCAGTGAGGCCAGAGGCCGGGCTGTGCTGGGGCCTGAGATGGGGTGGTGGGGAGAG
AGTCTCTCCCCTGCCCCTGTCTCTTCCGTGCAGGAGGAGCATGTTTAAGGGGAAGGGTTCAAAGCTGGTC
ACATCCCCAACAAAAAAGCCCACGGACAACGAAAAGCCCACTCGCTTGTCCAGTGCCACAGGAGGGGGCA
AGTGGAGGAGGAGAGGTGGCGGTGCTCCCCACTCCACTGCCAGTCGTCACTGGCTCTCCCTTCCCTTCAT
CCTCGTTCCCTATCTGTCACCATTTCCTGTCGTCGTTTCCTCTGAATGTCTCACCCTGCCCTCCCTGCTT
GCAAGTCCCCTGTCTGTAGCCTCACCCCTGTCGCATCCTGACTACAATAACAGCTTCTGGGTGTCCCCGG
CATCCACTCTCTCTCCCTTCTTATCCCTTCCGTGACGGATGCCTGAGGAACCTTCCCCAAACTCTTCTGT
CCCATCCCTGCCCTGCTCAAAATCCAATCACAGCTCCCTAACGCTCCTGAATCAACGTGAAGTCCTGTCT
TGAGTAATCCGTGGGCCCTAACTCACTCATCCCAACTCTTCACTCACTGCCTTGCCCCACACCCTGCCAG
```

After examining this DNA sequence, try answering the following questions:

1. Is this actually human DNA? If not, what organism is it from?
2. What is its biological function?
3. Is it going to kill you? (Hey, it could be poliovirus DNA. How would you know?)

My guess is that without a computer and bioinformatics, you don't stand a chance of answering these questions correctly. The above DNA sequence information codes for a small fragment of human DNA on chromosome 12. (Or it could be from a vampire bat. Keep reading to find out!) In fact, the entire human genome contains 1,000 times this much information. (BTW, this information is called sequence information because it is linked together as a sequence of letters.) And look how boring it is! The same 4 letters—A, G, C, and T—over and over again in different combinations. RNA and protein information looks pretty similar in the computer, except that one can tell RNA sequence data apart because it contains U instead of T. Protein sequence data are also easy to differentiate because up to 21 different letters representing the various amino acids are used in the sequences.

Here is some RNA sequence information:

```
GUUUAAGGGACACCGCAGAAAUGGUGAAUACAAUGAAGACAAAGCUGUUGUGUGUACUGCUGCUUUGTGG
```

And here is some protein sequence information:

```
RCDRGLAQCHTVPVKSCSELRCFNGGTCWQAASFSDFVCQCPKGYTGKQCEVDTHATCYKDQGVTYRGTW
```

Granted, the protein sequence is a little more interesting, but it is still pretty mind-numbing to stare at all day. However, mind-numbing tasks are exactly what computers were built for: determining the positions of all the known stars in our galaxy, calculating compound interest for billions of bank accounts, and searching through all the house cat video URLs on the Internet, among other things.

In fact, the amount of molecular sequence data has grown so vast, and the technologies for generating DNA sequence information from organisms have become so efficient, that computer processor and hard drive technologies are starting to fall behind the rate of biological information generation. The graph in Fig. 0.1 shows the exponential growth in the databanks from 1982 to 2008.

In 2008, the amount of information depicted in Fig. 0.1 was considered an extreme amount of information. Now a single researcher can, in a single day, generate DNA sequence information equivalent to all the total sequence information available in 2008.

So, what can one do with all these data? That question is the principal subject of this book, namely, how bioinformatics algorithms and banks of fancy computers can make sense of this growing mountain of molecular sequence data. In the next few sections you will learn a little about these critical biological molecules and how letters of the alphabet can be used to represent and store them in

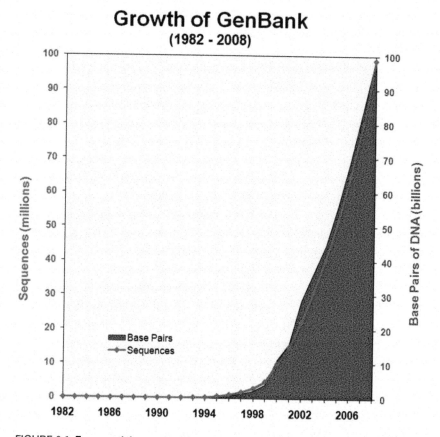

FIGURE 0.1. **Exponential growth of GenBank database from 1982 to 2008.** Courtesy of National Library of Medicine.

computers. Then, having provided a basic understanding of molecular biology and its computational representation, the rest of the book will focus on teaching you about the algorithms used to analyze these data.

DNA in the Computer

Deoxyribonucleic acid, otherwise (thankfully) known as DNA, is life's single most important molecule. DNA underpins virtually all of biology. With the exception of a few viruses,[1] life encodes itself using the chemical nucleotides of the DNA double helix. Every living cell contains its molecular information in the form of DNA, including the 1 trillion cells in the human body and every animal, plant, fungal, and bacterial cell on the planet. Placed end to end, the DNA from a single human cell could stretch an astonishing 3 meters.[2] The amount of raw information contained in this 3-meter length of DNA is similarly remarkable. The unit of molecular information in DNA is the nucleotide. Thus, the human genome, the complete set of all the DNA information contained in each cell, contains approximately 3 billion pieces of information.

Theoretically, the ability to read and interpret this chemical code should allow us to learn a great deal about how cells and organisms function and interact.

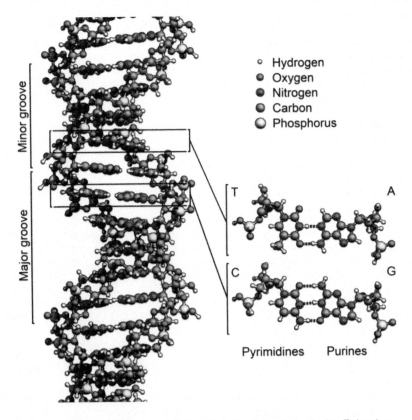

FIGURE 0.2. **Chemical structure of DNA at the atomic level.** A, adenine; T, thymine; C, cytosine; G, guanine. Courtesy of Zephyris (Richard Wheeler), under license CC BY-SA 3.0.

The chapters of this book show ways in which the combination of experimentation and DNA sequence analysis[3] can reveal powerful new insights into molecular processes, cellular mechanisms, disease, and biodiversity. However, before we can analyze DNA sequences in the computer, we must first store the DNA information in a computer. How do we represent the complex biochemical structure of DNA in a computer?

Figure 0.2 shows the chemical structure of DNA at the atomic level, showing all the individual atoms and bonds between them for just a small fraction of a typical DNA molecule. This figure of a DNA double helix shows the arrangement of individual atoms in a fragment of DNA. This DNA segment has a total of 14 nucleotide base pairs. Two of the base pairings are shown on the right side of the figure. A thymine (T) nucleotide binds to adenine (A), and cytosine (C) binds to guanine (G). If we were to store this structure in its full 3-dimensional (3D) glory, we would have to keep track of the position of every atom, bond, and bond angle for more than 60 atoms per DNA base pair. Remember, too, that the DNA in just one of your cells is approximately 300,000,000 times longer than the DNA in the figure.

Saving all these data, even in a big computer, is not very practical. Clearly, we need an alternative to storing every atom of every DNA molecule. Fortunately, the chemical structure of DNA is highly redundant and easy to simplify. Figure 0.3 shows that if we zoom in and flatten a piece of the structure, we can see the four nucleotides that make up DNA.

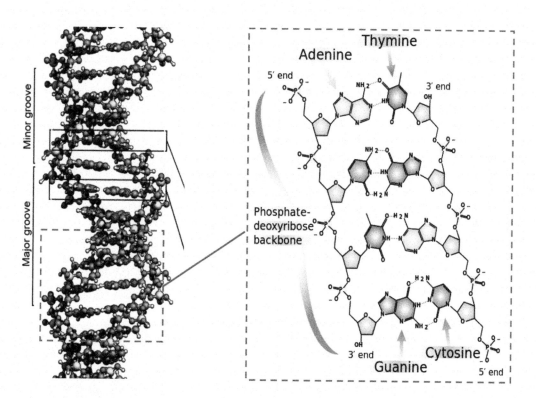

FIGURE 0.3. **Close-up view of a sequence of DNA showing chemical structures of the four nucleotide bases and their pairings.** Solid boxes correspond to those in Fig. 0.2. Courtesy of Zephyris (Richard Wheeler), under license CC BY-SA 3.0.

Each of the four nucleotides is composed of three parts: the phosphate that binds the nucleotides together, the deoxyribose sugar molecule, and the nucleoside base itself. The nucleoside base is really the only interesting bit. The DNA molecule is a long string of these four nucleotides base-paired to complementary nucleotides on the opposite strand. The box in Fig. 0.4 shows the A and T binding to one another, making a base pair. To make it simpler and easier to store in a computer, we can use single letters in place of the nucleotides: A, T, C, and G.

Single letters are perfect for data storage. The whole human genome is "only" 3 billion letters—that's just 6 MB of data, the size of a family photo, and you can store a lot of photos on a typical laptop. DNA data storage is even easier when you realize that one must store only half the data. DNA has two parallel strands running in opposite directions. In Fig. 0.4, the strand on the left runs from 5' (top) to 3' (bottom), while the rightmost strand runs in the opposite direction.

Since adenine ALWAYS binds thymine, and guanine ALWAYS binds cytosine, if you know one strand, you automatically know the other. If one stores the leftmost strand as ACTG, it is trivial to reconstruct the opposite strand as CAGT. The DNA nucleotides are always written left to right, from the 5' to 3' direction.[4] This is the sequence (the order) in which the nucleotides are written. When biologists talk about a DNA sequence this is what they are talking about. When you have one strand of a sequence, you can determine the other by finding its reverse complement. To do so, move in the reverse direction, from the end back to the beginning, and determine the matching base for each nucleotide.

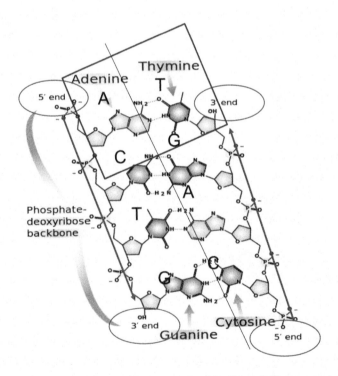

FIGURE 0.4. **Close-up view of base-paired nucleotides (boxed).** The arrows show how DNA strands run in opposite directions. A 5'-to-3' strand is paired with a complementary strand running in the 3' to 5' direction.

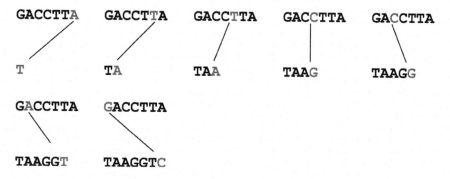

FIGURE 0.5. **Reverse complementation of a DNA sequence.**

For instance, say you have the DNA sequence GACCTTA. To reverse complement this sequence, go to the last letter, in this case A, and write the complementary base, T. Move backwards until you reach the beginning. An example is
shown in Fig. 0.5.

The final sequence is the reverse complement, which in this case is TAAGGTC.
Ta-da!

RNA in the Computer

DNA molecules are extremely exciting to the bioinformatician, but in the cell, DNA
is rather boring. DNA by itself does not catalyze any enzymatic reaction or perform any cellular activity. In order for cellular activity to occur, the nucleotide information of DNA needs to be "read" by the cell's complex cellular machinery
composed of proteins and RNA molecules. The way information passes from the
DNA to the cell is known as the central dogma of molecular biology, and it can be
summarized as follows:

DNA ⟶ [transcription] ⟶ mRNA ⟶ [translation] ⟶ protein

In the first step of this process, a protein enzyme complex that includes the RNA
polymerase unwinds the DNA double helix and transcribes one of the strands into
an RNA molecule called mRNA. The m in mRNA stands for messenger, because
the mRNA is a copy of the message contained in the DNA nucleotides.

RNA molecules are very similar to DNA but have a few key differences.

- RNA molecules are single-stranded.
- Instead of thymine, RNA molecules use uracil.
- RNA molecules contain a ribose sugar in the nucleotide instead of a deoxyri-
  bose sugar.

Figure 0.6 illustrates the basic process of transcription, in which the RNA polymerase protein complex makes a copy of one strand of the DNA double helix (the
coding strand) by synthesizing a complementary single-stranded RNA molecule.
Along the template strand, the RNA polymerase moves in a 3' to 5' direction, rec-
ognizing each DNA nucleotide, finding a complementary RNA nucleotide, and adding
this nucleotide to the 3' end of the growing RNA molecule. Once the synthesis is

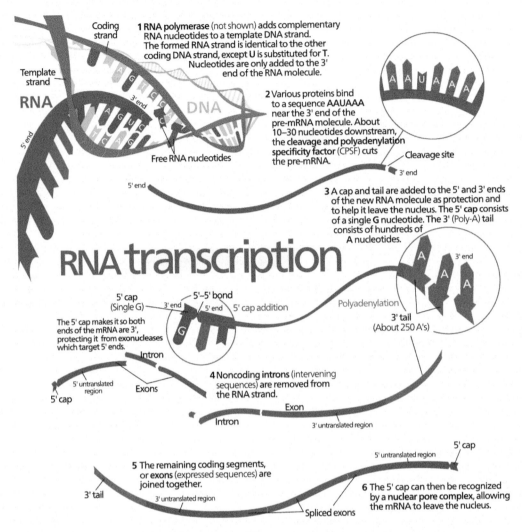

Coding strand

1 RNA polymerase (not shown) adds complementary RNA nucleotides to a template DNA strand. The formed RNA strand is identical to the other coding DNA strand, except U is substituted for T. Nucleotides are only added to the 3' end of the RNA molecule.

Template strand

RNA

3' end

DNA

5' end

Free RNA nucleotides

2 Various proteins bind to a sequence AAUAAA near the 3' end of the pre-mRNA molecule. About 10–30 nucleotides downstream, the **cleavage and polyadenylation specificity factor** (CPSF) cuts the pre-mRNA.

Cleavage site

3' end

5' end

3 A cap and tail are added to the 5' and 3' ends of the new RNA molecule as protection and to help it leave the nucleus. The 5' cap consists of a single G nucleotide. The 3' (Poly-A) tail consists of hundreds of A nucleotides.

RNA transcription

3' end

5' cap (Single G)

3' end

5'–5' bond

5' end 5' cap addition

Polyadenylation

3' tail (About 250 A's)

The 5' cap makes it so both ends of the mRNA are 3', protecting it from exonucleases which target 5' ends.

G

Intron

5' untranslated region

Exons

5' cap

4 Noncoding **introns** (intervening sequences) are removed from the RNA strand.

Exon

Intron

3' untranslated region

5' untranslated region

5' cap

5 The remaining coding segments, or **exons** (expressed sequences) are joined together.

3' tail 3' untranslated region

Spliced exons

6 The 5' cap can then be recognized by a **nuclear pore complex**, allowing the mRNA to leave the nucleus.

FIGURE 0.6. **Simplified illustration of transcription in eukaryotes (including humans).** Steps 1–6 describe the steps of the molecular process of transcription, in which the RNA polymerase protein complex (not shown) makes single-stranded RNA using one strand of the DNA double helix as a template. Step 1 (initiation and elongation) in the figure is common among all forms of life, while step 2 (termination) differs significantly between eukaryotes and bacteria. Steps 3–6 do not occur in nucleus-free bacteria. Courtesy of Kelvin Ma, under license CC BY 3.0.

complete, the RNA is uncoupled from the DNA, the RNA polymerase falls off, and the DNA double helix reforms.

A static image does not do justice to the beauty of transcription. Fortunately, the folks who run the DNA Learning Center (DNALC; **https://www.dnalc.org/**) have created stunning animations of many molecular processes. Here is the link to a 3D animated video of transcription:

https://www.dnalc.org/resources/3d/12-transcription-basic.html

A more advanced version can be found here:

https://www.dnalc.org/resources/3d/13-transcription-advanced.html

I highly encourage you to check out DNALC's collection of high-quality and beautiful animations.

Most of the RNA synthesized in the cell is eventually translated into a protein sequence, and as stated earlier, this type of RNA is called messenger RNA (mRNA) because it carries a message of information on how to synthesize the protein. Other types of RNA, known as structural or noncoding RNA, are also critical to cellular function but are not translated into proteins (see Chapter 05). The mRNA can be modified, shortened, or rearranged and eventually recycled by the cell without affecting the underlying DNA. Moreover, many RNAs can be synthesized from the same DNA template very quickly, one after another, and the cell controls the types and amounts of DNA made.

Protein Translation

The final destination of mRNA is the ribosome, the so-called protein factory of the cell. The ribosome is a macromolecular complex composed of two very large structural RNA molecules called ribosomal RNA (rRNA) and a lot of proteins. It is at the ribosome that the data encoded in the DNA is translated[5] into protein in a factory-like manner. Figure 0.7 shows a basic illustration of translation.

Again, the DNALC has created excellent translation videos that are very worth watching.

Basic protein translation animation:
https://www.dnalc.org/resources/3d/15-translation-basic.html

Advanced translation:
https://www.dnalc.org/resources/3d/16-translation-advanced.html

Protein Sequences in the Computer

Proteins are the molecular machines that make cells work, everything from copying DNA (the DNA polymerase enzyme) to gating molecules in and out of cells (sugars, ions, etc.) to storing molecular energy (ATP). From its earliest days, much of bioinformatics has focused on trying to predict the structure of proteins and other important properties, given their primary sequence (Fig. 0.8).

In the computer, the primary sequence of a protein is represented as a series of letters, each of which represents one of the 21 amino acids, and it represents the order that the amino acids are linked together end to end. Figure 0.9 illustrates the 21 amino acids, some of which are complicated and would be challenging to store in a computer.

The letters of the amino acids, like the four bases of DNA, are trivial to store in the computer. Furthermore, it is really easy to use computers to digitally translate DNA sequence to protein. The ease of creating protein translations in the computer from readily obtainable DNA sequences (cheap and fast!) means that computational algorithms are the primary way we learn information about protein sequences in

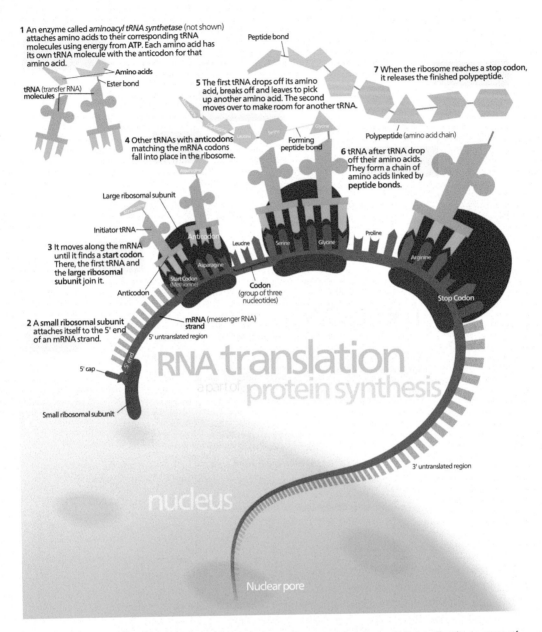

1 An enzyme called *aminoacyl tRNA synthetase* (not shown) attaches amino acids to their corresponding tRNA molecules using energy from ATP. Each amino acid has its own tRNA molecule with the anticodon for that amino acid.

Amino acids

Ester bond

tRNA (transfer RNA) molecules

Peptide bond

5 The first tRNA drops off its amino acid, breaks off and leaves to pick up another amino acid. The second moves over to make room for another tRNA.

7 When the ribosome reaches a stop codon, it releases the finished polypeptide.

4 Other tRNAs with anticodons matching the mRNA codons fall into place in the ribosome.

Forming peptide bond

Polypeptide (amino acid chain)

6 tRNA after tRNA drop off their amino acids. They form a chain of amino acids linked by peptide bonds.

Large ribosomal subunit

Initiator tRNA

3 It moves along the mRNA until it finds a **start codon**. There, the first tRNA and the large ribosomal subunit join it.

Anticodon

Asparagine

Asparagine

Leucine

Serine

Glycine

Proline

Arginine

Anticodon

Start Codon (Methionine)

Codon (group of three nucleotides)

Stop Codon

2 A small ribosomal subunit attaches itself to the 5' end of an mRNA strand.

mRNA (messenger RNA) strand

5' untranslated region

5' end

5' cap

Small ribosomal subunit

RNA translation

a part of protein synthesis

3' untranslated region

nucleus

Nuclear pore

FIGURE 0.7. Simplified illustration of translation. Steps 1–7 describe the process of protein translation in which a molecular "machine" called the ribosome translates mRNA into a protein. The steps shown in the figure are shared by all cellular life, though bacteria do not contain a nucleus. Courtesy of Kelvin Ma, under license CC BY 3.0.

new organisms. A new bacterial genome, for instance, can be sequenced and assembled in a day and yield 3,000+ protein sequences.[6]

For example, here is a portion of the DNA sequence that codes for the human hemoglobin beta protein, part of the complex that binds oxygen in red blood cells:

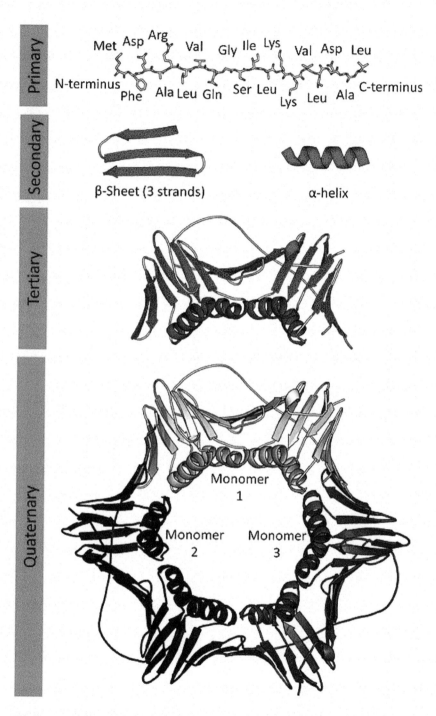

Primary

Met Asp Arg Val Gly Ile Lys Val Asp Leu
N-terminus   C-terminus
Phe Ala Leu Gln Ser Leu Lys Leu Ala

Secondary

β-Sheet (3 strands)   α-helix

Tertiary

Quaternary

Monomer 1

Monomer 2   Monomer 3

FIGURE 0.8. **Aspects of protein structure.** The structure of a protein can be described at four different levels. The primary structure is the sequence of bonded amino acids, unfolded. The amino acids fold into two basic types of structures known as secondary structures: alpha helices and beta sheets. The tertiary structure is the 3D structure of the protein and is the most important for understanding the protein's function. Finally, quaternary structure is an arrangement of multiple proteins bound together to make a single functional macromolecule. Image courtesy of Thomas Shafee, under license CC BY 4.0.

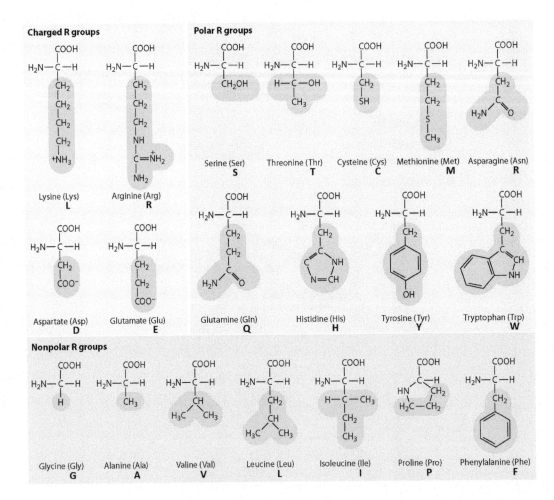

FIGURE 0.9. **Chemical structures of the 20 common amino acids that are used to make proteins.** The letters below the names of the amino acids can be used instead of the full names to represent the structure. For example, V represents valine, a hydrophobic amino acid.

ATGGTGCACCTGACTCCTGAGGAGAAGTCTGCCGTTACTGCCCTGTGGGGCAAGGTGAACGTGGATGAAG
TTGGTGGTGAGGCCCTGGGCAGGCTGCTGGTGGTCTACCCTTGGACCCAGAGGTTCTTTGAGTCCTTTGG
GGATCTGTCCACTCCTGATGCTGTTATGGGCAACCCTAAGGTGAAGGCTCATGGCAAGAAAGTGCTCGGT
GCCTTTAGTGATGGCCTGGCTCACCTGGACAACCTCAAGGGCACCTTTGCCACACTGAGTGAGCTGCACT
GTGACAAGCTGCACGTGGATCCTGAGAACTTCAGGCTCCTGGGCAACGTGCTGGTCTGTGTGCTGGCCCA
TCACTTTGGCAAAGAATTCACCCCACCAGTGCAGGCTGCCTATCAGAAAGTGGTGGCTGGTGTGGCTAAT
GCCCTGGCCCACAAGTATCAC

> Using the genetic code (Fig. 0.10) to computationally translate this sequence starting at the first base, here is the protein translation of this DNA sequence:

**MVHLTPEEKSAVTALWGKVNVDEVGGEALGRLLVVYPWTQRFFESFGDLSTPDAVMGNPKVKAHGKKVLG
AFSDGLAHLDNLKGTFATLSELHCDKLHVDPENFRLLGNVLVCVLAHHFGKEFTPPVQAAYQKVVAGVAN
ALAHKYH**

Second letter

	U	C	A	G	
U	UUU Phe UUC Phe UUA Leu UUG Leu	UCU Ser UCC Ser UCA Ser UCG Ser	UAU Tyr UAC Tyr UAA Stop UAG Stop	UGU Cys UGC Cys UGA Stop UGG Trp	U C A G
C	CUU Leu CUC Leu CUA Leu CUG Leu	CCU Pro CCC Pro CCA Pro CCG Pro	CAU His CAC His CAA Gln CAG Gln	CGU Arg CGC Arg CGA Arg CGG Arg	U C A G
A	AUU Ile AUC Ile AUA Ile AUG Met	ACU Thr ACC Thr ACA Thr ACG Thr	AAU Asn AAC Asn AAA Lys AAG Lys	AGU Ser AGC Ser AGA Arg AGG Arg	U C A G
G	GUU Val GUC Val GUA Val GUG Val	GCU Ala GCC Ala GCA Ala GCG Ala	GAU Asp GAC Asp GAA Glu GAG Glu	GGU Gly GGC Gly GGA Gly GGG Gly	U C A G

First letter   Third letter

FIGURE 0.10. Genetic code for eukaryotes, like us. The possible triplet codons of mRNA are listed with the amino acids they encode in eukaryotes (bacteria have a slightly different genetic code). In this table, the AUG start codon is shaded in green. The stop codons, which cause termination of translation, are shaded in pink.

Each group of three nucleotides in the DNA sequence codes for a different amino acid in the protein sequence. For example, the first three nucleotides, ATG, code for M (methionine). In Fig. 0.10, the table shows the RNA sequences which correspond to particular amino acids, but one can also use the DNA nucleotides. To do this: (i) transcribe the DNA into RNA (easy), (ii) break up the RNA sequence into groups of three nucleotides (i.e., codons), and (iii) match the codons to the amino acids using the table. For example, the RNA codon CUU codes for leucine (Leu), which is stored as an L in the computer.

The Molecular Structure of a Gene

The gene is the basic unit of heredity that determines some aspect of the organism. Genetic information is, of course, encoded in the organism's DNA, and this information determines how a particular polypeptide (protein) or nucleotide polymer (RNA) is produced. However, the DNA nucleotides in the genome that directly encode a protein make up only a small portion of the gene, particularly in eukaryotes. Eukaryotes encompass all multicellular organisms, including all animals, plants, and fungi, as well as many single cellular organisms, such as yeasts, parasites, and algae. Genes in eukaryotes can be extremely complex, and this complexity allows eukaryotes to (i) develop diverse body plans using nearly identical

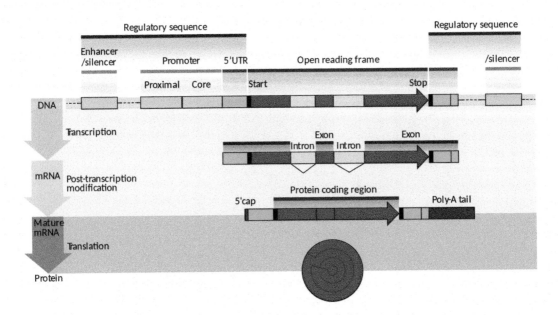

FIGURE 0.11. **General structure of a eukaryotic gene.** Only approximately 5% of human and other multicellular eukaryote DNA actually codes for proteins. This complexity allows for highly differentiated gene regulation, unique for different cell types (e.g., brain cells versus kidney cells), and the ability to make multiple different proteins from one protein-coding region. Transcription is controlled by proteins called transcription factors, which influence the binding of RNA polymerase either positively (more transcription) or negatively (less transcription). Transcription factors bind thousands of bases upstream in enhancer/silencer regions and also very close by in the proximal promoter regions. The RNA polymerase complex binds at the core promoter region and acts to transcribe what is known as the pre-mRNA, which includes regions termed exons and introns. Regulatory regions 3' of the final exon signal the RNA polymerase to stop transcription. The pre-mRNA is further processed by cellular proteins (posttranslational modification) prior to translation. This processing includes adding a 5' cap and a series of adenine nucleotides at the end of the mRNA [called the poly(A) tail] followed by the process of splicing, in which the intron regions are removed and the exons spliced together. The splicing process allows multiple different proteins to be produced from one pre-mRNA transcript. For instance, in the figure the middle exon could be removed, making a shorter protein. While most genes have only one or two splice variants, 1 to 2% of human genes have nine or more alternative splice variants. Courtesy of Thomas Shafee, under license CC BY 4.0.

proteins, (ii) differentially regulate protein and RNA production in thousands of cell types in different tissues, and (iii) create many combinations of proteins from the same gene.

In order to better understand the purpose of bioinformatics algorithms and databases, it is important to have some comprehension of the structural and functional elements of genes. Figure 0.11 diagrams the standard elements in a eukaryotic gene. Eukaryotic genes can be many thousands of nucleotide bases long when all aspects are accounted for, and only a small portion is devoted to encoding the protein sequence. The rest is devoted to binding proteins that regulate transcription and translation.

Figure 0.12 describes the structure of a bacterial gene operon. Single-celled bacteria do not have complex cell types or body plans, so their genome organiza-

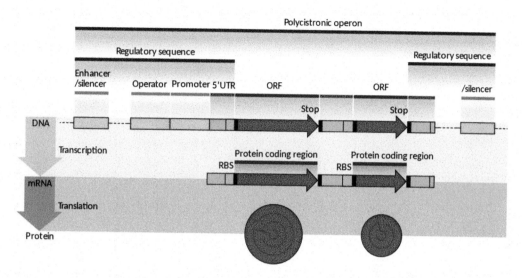

FIGURE 0.12. **General structure of a bacterial gene.** The regulatory structure of a bacterial gene is similar to but much simpler than that in eukaryotes. Transcription and its regulation are controlled by far fewer transcription factors. The other main differences are the lack of introns and the fact that mRNAs for most coding regions (ORFs, which stands for open reading frames) are transcribed in clusters. For instance, the mRNAs for all the protein enzymes involved in making the amino acid tryptophan are synthesized and ultimately transcribed at the same time, a very efficient process in busy bacteria. Note that not all bacterial genes have enhancers/silencers. UTR, untranslated region; RBS, ribosome binding site. Courtesy of Thomas Shafee, under license CC BY 4.0.

tion is much simpler. However, as single cells, bacteria must adapt quickly to environmental conditions, producing metabolic enzymes, proteins involved in motility, or cell surface proteins very rapidly. Thus, unlike eukaryotes, bacteria cluster their genes into operons, contiguous regions of DNA that each code for a different polypeptide (protein) involved in the same functional process. For example, all the proteins involved in metabolizing glucose are next to each other in an operon. In bacteria, the process of transcription is directly linked to translation, and they often happen more or less simultaneously. In eukaryotes, mRNA is exported outside the nucleus of the cell for translation.

Notes

1. Some viruses, like HIV and influenza virus, use RNA (the chemical cousin of DNA) as their genetic material.
2. It would be rather thin, though, just 20 nanometers in diameter, or 5,000 times thinner than a human hair.
3. Also the analysis of RNA and protein sequences, which can be determined from DNA sequences.
4. The protein enzymes that bind to DNA to (biochemically) read it, or copy it, recognize directionality and always move in the 3' to 5' direction on the DNA.
5. A note on transcription and translation. Transcription is the process of copying the same text from one place to another. The mRNA is pretty much the same language (nucleotides). Translation is the process of expressing words in another language. DNA and RNA sequences are like one chemical language, and protein sequences are like another.
6. Hence another need for bioinformatics. Imagine the number of experiments that would be needed to test all the functions of these proteins at the bench!

ACTIVITY 0.1   BIOLOGICAL DATABASES AND DATA STORAGE

Motivation

The goals of this section are to help you (i) understand why computers are increasingly vital in the biological sciences, (ii) learn the basics of how biological information, namely, DNA, RNA, and protein sequence information, is stored in the computer, and (iii) be able to interpret common data file types used for storing biological information. DNA is life's principal information storage device and is present in every living cell. Some viruses, like Ebola, use RNA, but are they really "alive"? (Statement intended to annoy virologists.) Complete understanding of an organism's DNA tells you everything about the organism: what proteins it makes, when and how it activates genes, how different cell types (e.g., heart, liver, and brain cells) develop, and more.

However, because researchers generate so much DNA sequence from so many organisms, and it is so very boring to look at, we need computers to make any sense of these data. Below is a sequence of human DNA, with each letter representing a nucleotide in the DNA. This sequence is a tiny fraction of the 3 BILLION nucleotides in the human genome:

```
CTGATGGGAATGCAAGCAGCCATTGAGCAGGCTATGAAGAGTCGTGAGATTCTGGGCATCTCAGACCCTC
AGACGCTGGCCCATGTGCTGACAGCCGGAGTGCAGAGTTCCTTGAATGACCCACGCCTCTTCATCTCCTA
TGAGCCCAGTACCCTCGAGGCTCCCCAGNCAGCACCAACACTCACCAACCTCACCCGAGAAGAACTACTG
GCCCAGCTACAGAGGAGCATCCACCATGAGGTCCTTGAGGGCAACGTGGGTTACCTACGAATAGATGATT
TCCCCGGCCAGGAGGTACTGAGTGAGCTGGGGGGATTCTTGGTGACCCATATGTGGAGGCAGCTCATGGA
CACCTCCTCCTTGGTGCTCGATCTCCGGTACTGTGCTGGTGGTCACATCTCTGGGATCCCTTATTTCATC
```

Note how this sequence is just the same four letters over and over in different combinations—terribly dull to read and nearly impossible to interpret. However, with fancy computational algorithms and banks of computers, we could use these letters to determine in seconds the species that the sequence came from, whether the sequence codes for a protein, structural RNA, or a piece of ancient "junk" DNA, and the cellular role of the gene coded for by this DNA. (This one is a piece of Florida muskrat DNA that codes for an eye gene.) Far out, right?

Before discussing how one computationally analyzes DNA, RNA, or protein sequences, one first needs to understand the ways in which these sequences are stored in the computer. After covering the basics of biological sequences in the introduction, the exercises will cover the National Center for Biotechnology Information's (NCBI) scientific journal article search engine and storage database (PubMed) and the underlying text format (MEDLINE). The exercises and tutorials will then cover the complex, information-rich GenBank data file format and the much simpler FASTA file format. Finally, we will briefly review a series of very useful biological information databases for exploring protein sequences and microbial and eukaryotic genomes.

Learning Objectives

1. Understand how and why biological information is stored electronically (Motivation).
2. Gain familiarity with some commonly used data storage formats and how to interpret them (Concepts and Exercises).

3.   Learn the basics of some powerful and highly useful biological sequence databases (Lab Exercises).

Concepts

To develop an appreciation of the types of data that can be stored in biological databases, use your knowledge of biology and cleverness to determine what the elements shown in boldface below indicate about a particular DNA sequence stored in a database at NCBI. The following text file is an example of a GenBank formatted file. Write your guess as to what the items indicated in boldface mean in the nearest box.

```
LOCUS         HSDUT2                    1177 bp    DNA       linear   PRI 28-SEP-1997
DEFINITION    Homo sapiens dUTPase (DUT) gene, exon 3.
ACCESSION     AF018430
VERSION       AF018430.1  GI:2443576
KEYWORDS      .
SEGMENT       2 of 4
SOURCE        Homo sapiens (human)
  ORGANISM    Homo sapiens
              Eukaryota; Metazoa; Chordata; Craniata; Vertebrata; Euteleostomi;
              Mammalia; Eutheria; Euarchontoglires; Primates; Haplorrhini;
              Catarrhini; Hominidae; Homo.
REFERENCE     1  (bases 1 to 1177)
  AUTHORS     Pearlman,R.E.
  TITLE       Human genomic nuclear and mitochondria dUTPase gene
  JOURNAL     Unpublished
REFERENCE     2  (bases 1 to 1177)
  AUTHORS     Pearlman,R.E.
  TITLE       Direct Submission
  JOURNAL     Submitted (11-AUG-1997) Biology, York University, 4700 Keele St.,
              North York, ONT M3J 1P3, Canada
FEATURES             Location/Qualifiers
     source          1..1177
                     /organism="Homo sapiens"
                     /mol_type="genomic DNA"
                     /db_xref="taxon:9606"
                     /map="15q15-q21.1"
     gene            order(AF018429.1:<1..1735,1..1177,AF018431.1:1..45,
                     AF018432.1:658..732,AF018432.1:884..954,
                     AF018432.1:1391..>1447)
                     /gene="DUT"
     mRNA            join(AF018429.1:<282..561,AF018429.1:1034..1172,560..651,
                     AF018431.1:1..45,AF018432.1:658..732,AF018432.1:884..954,
                     AF018432.1:1391..>1447)
                     /gene="DUT"
                     /product="dUTPase"
                     /note="alternatively spliced;
                     encodes mitochondrial form
                     of the protein"
     CDS             join(AF018429.1:282..561,AF018429.1:1034..1172,560..651,
                     AF018431.1:1..45,AF018432.1:658..732,AF018432.1:884..954,
                     AF018432.1:1391..1447)
                     /gene="DUT"
```

```
                    /note="DUT-M; alternatively spliced; mitochondrial form of
                    the protein; similar to H. sapiens dUTPase encoded by
                    GenBank Accession Number U90224"
                    /codon_start=1
                    /product="dUTPase"
                    /protein_id="AAB71393.1"
                    /db_xref="GI:2443580"
                    /translation="MTPLCPRPALCYHFLTSLLRSAMQNARGTAEGRSRGTLRARPAP
                    RPPAAQHGIPRPLSSAGRLSQGCRGASTVGAAGWKGELPKAGGSPAPGPETPAISPSK
                    RARPAEVGGMQLRFARLSEHATAPTRGSARAAGYDLYSAYDYTIPPMEKAVVKTDIQI
                    ALPSGCYGRVAPRSGLAAKHFIDVGAGVIDEDYRGNVGVVLFNFGKEKFEVKKGDRIA
                    QLICERIFYPEIEEVQALDDTERGSGGFGSTGKN"
     mRNA           join(AF018429.1:<1018..1172,560..651,AF018431.1:1..45,
                    AF018432.1:658..732,AF018432.1:884..954,
                    AF018432.1:1391..>1447)
                    /gene="DUT"
                    /product="dUTPase"
                    /note="alternatively spliced; encodes nuclear form of the
                    protein"
     CDS            join(AF018429.1:1018..1172,560..651,AF018431.1:1..45,
                    AF018432.1:658..732,AF018432.1:884..954,
                    AF018432.1:1391..1447)
                    /gene="DUT"
                    /note="DUT-N; alternatively spliced; nuclear form of the
                    protein; similar to H. sapiens dUTPase encoded by GenBank
                    Accession Number U90224"
                    /codon_start=1
                    /product="dUTPase"
                    /protein_id="AAB71394.1"
                    /db_xref="GI:2443581"
                    /translation="MPCSEETPAISPSKRARPAEVGGMQLRFARLSEHATAPTRGSAR
                    AAGYDLYSAYDYTIPPMEKAVVKTDIQIALPSGCYGRVAPRSGLAAKHFIDVGAGVID
                    EDYRGNVGVVLFNFGKEKFEVKKGDRIAQLICERIFYPEIEEVQALDDTERGSGGFGS
                    TGKN"
     exon           560..651
                    /gene="DUT"
                    /number=3
ORIGIN
        1 tccctaaatc aacacagatc atgtggagga ataaaatggg gttaatatat gtaaaaccaa
       61 ttaggaaact gtttctgggg caacacagta aagggcttat tcaatggata ggctagtatt
      121 attagttagt aattgggccc ttttttttctt tgtttctttt cttcattttt ttccttttca
      181 aactatgggt tgtaaagcat ccaccttttg aaagtttgcc tttctgccct ttcacgctga
      241 taagtacctc agtttccaat aaacttttgt tcaggggcaa acatttacaa tgttgacatc
      301 tcttcacacc accaaaaata ttcatggaga attattttat ctaaagctgt cttttttaata
      361 ataaaatagc cacctctacc ttcttcataa actttttaaga tgaattggta attcatcata
      421 gcaaggttga ttttagaaac taaagttgca ttaattcatt aaatacactg aaagtaattt
      481 tgtatgcttg gtcacaaaga aaatataaaa acaattttat aaatagattt gcagttattt
      541 tctttcaata ttttcttagt gcctatgatt acacaatacc acctatggag aaagctgttg
      601 tgaaaacgga cattcagata gcgctcccct ctgggtgtta tggaagagtg ggtaagtcat
      661 ttaagaaaca ggtaactatt tgtcaagttc tcctttgtga tagattcttc atgtttcatt
      721 tggggtaata agcaggcaat attgcttggg ctgtgtccta aaagaagcac catttgtgat
      781 agcaaatgca ctctttgaaa ggctttattt acatctctgc tttgcctctt tttgaccctt
      841 ttattttct cctcctcac tggagctttt aggctcacac tggcctagaa ggctgttctc
      901 agaacatggc attttatatt atgagagtaa aacttctgac ctgttggtcc cagaatgtgt
```

```
 961 aagcctactt aaccttttct tgtttggcca tggggtttag ggtaagggat actcttcagt
1021 gtttgtagag gcactgggag gaagctagga caaaatggag ttacacgtca acaggtttga
1081 tttttcctgg aagcgaattc agtgtttacc agacagttcc tttgcagagc gttagttcct
1141 ttttgactac ttccaagtta acttaaggag gcatgga
//
```

The answers are shown below. This file format can be quite rich in information. Indeed, one can learn a lot about the gene function, splicing, regulation, and other things by looking at such files.

```
LOCUS       HSDUT2                   1177 bp    DNA      linear   PRI 28-SEP-1997
DEFINITION  Homo sapiens dUTPase (DUT) gene, exon 3.
ACCESSION   AF018430
VERSION     AF018430.1 GI:2443576
KEYWORDS    .
SEGMENT     2 of 4
SOURCE      Homo sapiens (human)
  ORGANISM  Homo sapiens
            Eukaryota; Metazoa; Chordata; Craniata; Vertebrata; Euteleostomi;
            Mammalia; Eutheria; Euarchontoglires; Primates; Haplorrhini;
            Catarrhini; Hominidae; Homo.
REFERENCE   1 (bases 1 to 1177)
  AUTHORS   Pearlman,R.E.
  TITLE     Human genomic nuclear and mitochondria dUTPase gene
  JOURNAL   Unpublished
REFERENCE   2 (bases 1 to 1177)
  AUTHORS   Pearlman,R.E.
  TITLE     Direct Submission
  JOURNAL   Submitted (11-AUG-1997) Biology, York University, 4700 Keele St.,
            North York, ONT M3J 1P3, Canada
FEATURES             Location/Qualifiers
     source          1..1177
                     /organism="Homo sapiens"
                     /mol_type="genomic DNA"
                     /db_xref="taxon:9606"
                     /map="15q15-q21.1"
     gene            order(AF018429.1:<1..1735,1..1177,AF018431.1:1..45,
                     AF018432.1:658..732,AF018432.1:884..954,
                     AF018432.1:1391..>1447)
                     /gene="DUT"
     mRNA            join(AF018429.1:<282..561,AF018429.1:1034..1172,560..651,
                     AF018431.1:1..45,AF018432.1:658..732,AF018432.1:884..954,
                     AF018432.1:1391..>1447)
                     /gene="DUT"
                     /product="dUTPase"
                     /note="alternatively spliced;
                     encodes mitochondrial form
                     of the protein"
```

A unique code specific to this particular GenBank entry.

The number of nucleotides in this file. (Or amino acids if this was a protein entry.) See below.

Indicates the exon positions in a splice variant. Note: some of the exons are in other GenBank entries.

```
     CDS            join(AF018429.1:282..561,AF018429.1:1034..1172,560..651,
                    AF018431.1:1..45,AF018432.1:658..732,AF018432.1:884..954,
                    AF018432.1:1391..1447)
                    /gene="DUT"
                    /note="DUT-M; alternatively spliced; mitochondrial form of
                    the protein; similar to H. sapiens dUTPase encoded by
                    GenBank Accession Number U90224"
                    /codon_start=1
                    /product="dUTPase"
                    /protein_id="AAB71393.1"
                    /db_xref="GI:2443580"
                    /translation="MTPLCPRPALCYHFLTSLLRSAMQNARGTAEGRSRGTLRARPAP
                    RPPAAQHGIPRPLSSAGRLSQGCRGASTVGAAGWKGELPKAGGSPAPGPETPAISPSK
                    RARPAEVGGMQLRFARLSEHATAPTRGSARAAGYDLYSAYDYTIPPMEKAVVKTDIQI
                    ALPSGCYGRVAPRSGLAAKHFIDVGAGVIDEDYRGNVGVVLFNFGKEKFEVKKGDRIA
                    QLICERIFYPEIEEVQALDDTERGSGGFGSTGKN"
     mRNA           join(AF018429.1:<1018..1172,560..651,AF018431.1:1..45,
                    AF018432.1:658..732,AF018432.1:884..954,
                    AF018432.1:1391..>1447)          The protein sequence translation
                    /gene="DUT"                      for a splice mRNA.
                    /product="dUTPase"
                    /note="alternatively spliced; encodes nuclear form of the
                    protein"
     CDS            join(AF018429.1:1018..1172,560..651,AF018431.1:1..45,
                    AF018432.1:658..732,AF018432.1:884..954,
                    AF018432.1:1391..1447)
                    /gene="DUT"
                    /note="DUT-N; alternatively spliced; nuclear form of the
                    protein; similar to H. sapiens dUTPase encoded by GenBank
                    Accession Number U90224"
                    /codon_start=1
                    /product="dUTPase"
                    /protein_id="AAB71394.1"
                    /db_xref="GI:2443581"
                    /translation="MPCSEETPAISPSKRARPAEVGGMQLRFARLSEHATAPTRGSAR
                    AAGYDLYSAYDYTIPPMEKAVVKTDIQIALPSGCYGRVAPRSGLAAKHFIDVGAGVID
                    EDYRGNVGVVLFNFGKEKFEVKKGDRIAQLICERIFYPEIEEVQALDDTERGSGGFGS
                    TGKN"
     exon           560..651
                    /gene="DUT"
                    /number=3
ORIGIN
        1 tccctaaatc aacacagatc atgtggagga ataaaatggg gttaatatat gtaaaaccaa
       61 ttaggaaact gtttctgggg caacacagta aagggcttat tcaatggata ggctagtatt
      121 attagttagt aattgggccc ttttttttctt tgtttctttt cttcattttt ttcctttttca
      181 aactatgggt tgtaaagcat ccaccttttg aaagtttgcc tttctgccct ttcacgctga
      241 taagtacctc agtttccaat aaacttttgt tcaggggcaa acatttacaa tgttgacatc
      301 tcttcacacc accaaaaata ttcatggaga attattttat ctaaagctgt ctttttaata
      361 ataaaatagc cacctctacc ttcttcataa acttttaaga tgaattggta attcatcata
      421 gcaaggttga ttttagaaac taaagttgca ttaattcatt aaatacactg aaagtaattt
      481 tgtatgcttg gtcacaaaga aaatataaaa acaattttat aaatagattt gcagttattt
      541 tctttcaata ttttcttagt gcctatgatt acacaatacc acctatggag aaagctgttg
      601 tgaaaacgga cattcagata gcgctccctt ctgggtgtta tggaagagtg ggtaagtcat
      661 ttaagaaaca ggtaactatt tgtcaagttc tcctttgtga tagattcttc atgtttcatt
```

```
 721 tggggtaata agcaggcaat attgcttggg ctgtgtccta aaagaagcac catttgtgat
 781 agcaaatgca ctctttgaaa ggctttattt acatctctgc tttgcctctt tttgacccctt
 841 ttatttttct ccttcctcac tggagctttt aggctcacac tggcctagaa ggctgttctc
 901 agaacatggc attttatatt atgagagtaa aacttctgac ctgttggtcc cagaatgtgt
 961 aagcctactt aacctttct tgtttggcca tggggtttag ggtaagggat actcttcagt
1021 gtttgtagag gcactgggag gaagctagga caaaatggag ttacacgtca acaggtttga
1081 tttttcctgg aagcgaattc agtgtttacc agacagttcc tttgcagagc gttagttcct
1141 ttttgactac ttccaagtta acttaaggag gcatgga
```
//

The nucleotide sequence of the GenBank entry.

Exercises

Lab exercises (practice)

In this part of the exercise, you will learn the basics of some very useful biological databases. Follow the link or QR code below to background information on data formats and tutorials on databases you should use to answer the lab exercise questions. The Basics link includes tutorials for the UniProt (protein) and Ensembl (genome) databases, and the large collection of databases available at NCBI. The NCBI drop-down menu includes links to at least 45 different databases, allowing analysis of everything from gene expression to biochemistry to taxonomy.

From the Basics Link, under the Databases heading, you can click to learn about the NCBI and UniProt databases. After reviewing these, complete the Lab Exercise on the next page.
http://kelleybioinfo.org/algorithms/basics/

Lab Exercise

Part 1: using NCBI PubMed

1. How would you search in PubMed for all papers about bacteria written by Pace AND Lane? Use the MEDLINE author field to be more specific. Write the PubMed search terms you used here.

2. Use the MEDLINE entry of the Pace and Lane article titled "Evolutionary relationships among sulfur- and iron-oxidizing eubacteria" to get the GenBank accession numbers in the article. Write the first five accession numbers below.

3. How would you find all the experts on progesterone in San Diego? Do a specific search for the keyword and address. Write the search terms you would use below.

4. Find the PubMed PMID for any two papers from authors at San Diego State University that have worked with 16S. Write the search term used and the PMIDs below.

   Search Terms:

   PMID 1:

   PMID 2:

5. Use a combination of the search field and the left sidebar to search for review papers about *Mycobacterium* published in the last 5 years. Write the title of the first search result below:

   Title:

Part 2: data formats

1. Write the title line for the FASTA entry below for the DNA sequence of the nuclear splice mRNA variant description from GenBank entry AH005568.2.

   What is a nuclear splice mRNA variant anyway? Explain briefly below. (Hint: Ask Dr. Wikipedia or Professor Google for things you don't know or recognize. The Internet: not just for cat videos anymore!)

2. Find the following information for the human estrogen receptor (accession: NM_000125):

   a. What is the sequence of the polyadenylation [poly(A)] signal? Hint: look for "polyA_signal_sequence" in the file.

   b. Write the title line of the coding region protein sequence in FASTA format of estrogen receptor alpha isoform 1 (*Homo sapiens* estrogen receptor 1):

   c. At what position of the DNA record does the protein coding region start for the alpha isoform?

   d. What is a poly(A) signal anyway (i.e., what is its function)?

Part 3: microbial genomes and UniProt databases

1. Search the NCBI Genomes for *Sulfolobus solfataricus* P2.

   a. What kind of archaeon is *Sulfolobus solfataricus* (next term in the lineage after *Archaea*)?

   b. What does it metabolize?

   c. What is this microbe's optimal growth temperature?

   d. At what pH range does it grow best?

   e. What is the number of currently predicted proteins in the genome?

   f. Find the FASTA protein sequence of 30S ribosomal protein S3. Hint: click on the protein link (part e) and search the protein table for "30S ribosomal protein S3." Then select the link under "Protein product." Write the title line of the FASTA entry below.

2. Using UniProt, find the following information about the gene involved in cystic fibrosis transmembrane conductance regulation in *Homo sapiens*.

   a. What is the accession number of the full-length entry? (No fragments.)

   b. Describe briefly the function of this gene (see database entry under Function).

   c. What is its tissue specificity?

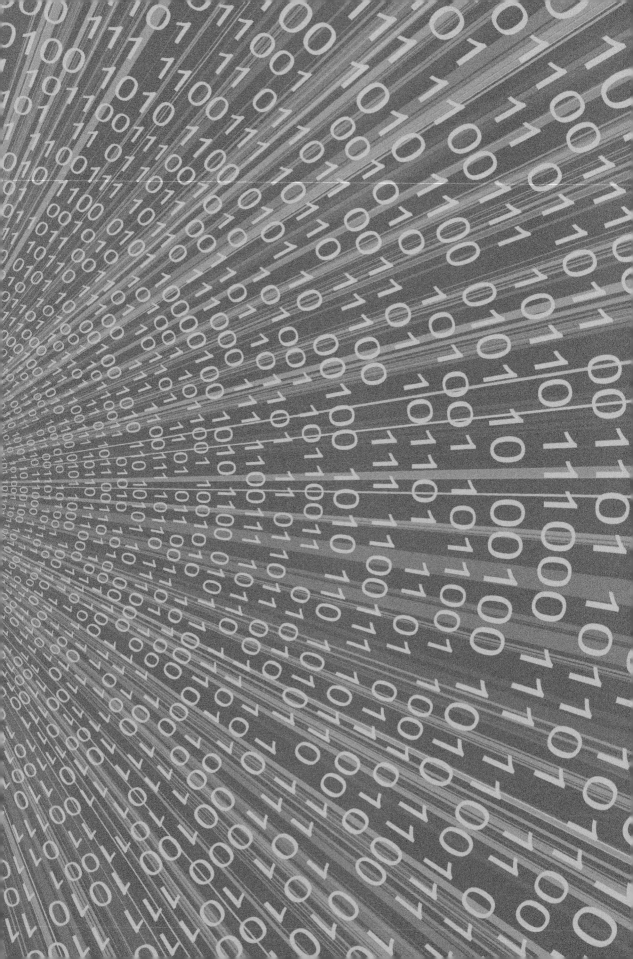

BLAST

The inventors of the BLAST (Basic Local Alignment Search Tool) computer program should win the Nobel Prize in Physiology or Medicine. I'm serious! Actually, they should probably share it with the inventors of the FASTA algorithm, which preceded the more efficient and flexible BLAST algorithm.

Why should the makers of a computer program that finds the best database match to a biological sequence win the most prestigious award in biomedical science? The Nobel Prize is usually reserved for discoveries made in the laboratory, such as the discovery of novel disease organisms, how to genetically engineer mice, proteins that glow in the dark, the mechanism of cell death, that sort of thing. A computer program seems a paltry gimmick in comparison, and biologists often chuckle or even scoff when I suggest that it deserves the Nobel. Determining the best match of a biological sequence appears rather uninspiring compared with the discovery of antibiotics or proteins that act like viruses. Yet in an era of rapidly expanding DNA sequencing capability and big biological data, BLAST's simplicity, elegance, and efficiency make this algorithm the most powerful biomedical discovery tool in the history of science.

BLAST It

Researchers use BLAST so often that it has been turned into a verb. (Like "to google" or "to chill".) "Hey, did you BLAST that sequence I gave you yet? No? Well, what are you waiting for? BLAST it!" To understand why BLAST is so useful and popular, imagine that you work in a molecular biology lab studying a novel infectious disease and you just received the results of your very first sequencing run, perhaps the first genetic information in the history of this unknown organism. Unfortunately, the sequence looks like this[1]:

```
>FX093345
TGCAGTCGATCATCAGCATACCTAGGTTTCGTCCGGGTGTGACCGAAAGGTAAGATGGAGAGCCTTGTTC
TTGGTGTCAACGAGAAAACACACGTCCAACTCAGTTTGCCTGTCCTTCAGGTTAGAGACGTGCTAGTGCG
TGGCTTCGGGGACTCTGTGGAAGAGGCCCTATCGGAGGCACGTGAACACCTCAAAAATGGCACTTGTGGT
CTAGTAGAGCTGGAAAAAGGCGTACTGCCCCAGCTTGAACAGCCCTATGTGTTCATTAAACGTTCTGATG
CCTTAAGCACCAATCACGGCCACAAGGTCGTTGAGCTGGTTGCAGAAATGGACGGCATTCAGTACGGTCG
TAGCGGTATAACACTGGGAGTACTCGTGCCACATGTGGGCGAAACCCCAATTGCATACCGCAATGTTCTT
CTTCGTAAGAACGGTAATAAGGGAGCCGGTGGTCATAGCTATGGCATCGATCTAAAGTCTTATGACTTAG
```

That's some exciting research you have there! Well, maybe, but what exactly is it? Clearly, it is a DNA sequence of some sort, but is it the secret to uncovering a deadly new pathogen, or is it human DNA contamination from the technician at the sequencing facility? Is it part of a protein toxin that allows a deadly microbe to destroy epithelial cells in the lungs, or is it a regulatory gene sequence from a housefly that landed on your pipette tip when you weren't looking?

Eyeballing a bunch of letters, the same four letters (A, G, T, and C) repeated in different orders, will get you nowhere. Wouldn't it be great if, without pilfering a pipette or pouring a petri plate, you could answer the following questions?

- Is the genetic material from a bacterium or a virus?
- Did you accidentally sequence a contaminating organism?
- Does the sequence code for a protein?
- What is the biological function of the sequence?
- Is the genetic material related to other, possibly well-characterized, pathogens?
- Is the sequence new to science?

Using BLAST, you can answer most of these questions in milliseconds. That's how long it takes for BLAST and a bunch of remote supercomputers at the National Center for Biotechnology Information (NCBI) to match your sequence to every DNA sequence ever catalogued. Figure 1.1 shows the BLAST results for your hypothetical new pathogen. In under a second, you discovered (i) the nature of the organism, namely, a known deadly virus of the coronavirus (a common cold virus) family, (ii) that your sequence encodes a protein, and (iii) that the protein is part of the virus's outer shell, a critical part of the infectious process. Not bad for a few seconds of effort. (Your tax dollars at work!)

Scaling Up: Massive Parallelization of BLAST

Imagine that instead of just one mystery sequence you had 10, 100, or 1,000 sequences and you had to match them against databases containing millions of sequences. Furthermore, these sequences encompass multiple different organisms, are part of different molecular processes, and are involved in a variety of cellular functions. Now you begin to see the scale of the problem and why bioinformatics can be so darn useful. By running BLAST searches in parallel on a suite of high-performance computers, it is possible to analyze thousands of mystery sequences against millions of known sequences generated by researchers all over the world (Fig. 1.2).

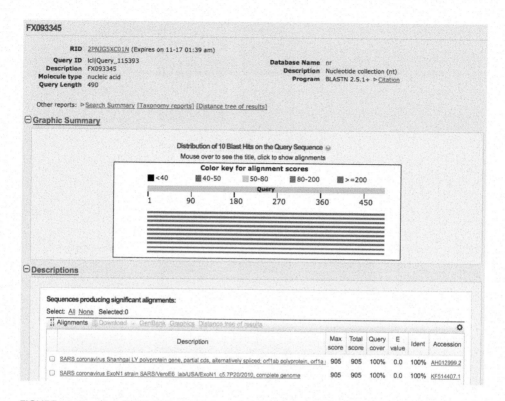

FIGURE 1.1. Results of a BLAST nucleotide search with the FX093345 nucleotide sequence. The results, returned in under a second, reveal that the sequence is an exact match to the genome of a severe acute respiratory syndrome (SARS) coronavirus protein first isolated in Shanghai (apparently misspelled "Shanhgai" in the database), China. The first 50 matches were to SARS coronaviruses isolated at different times, which provides confidence that we have indeed isolated a SARS virus. At the time this BLAST analysis was performed, the sequence was a perfect 100% match (bottom right under the "Ident" column) to more than one sequence, so we cannot say with certainty that it is the Shanghai strain. The fact that so many were identical also means that the results appear in a random order. That, and the fact that databases are constantly updated with new sequences, means that another search with the sequence may return different results from those seen in the figure. One of the most important considerations for interpreting BLAST results is the E-value. The E-value is the Expectation value, which tells the user how likely the search similarity result is due to random chance. An E-value of 1e-4 is the same as 0.0001, which indicates this similarity would occur 1 in 10,000 times by chance. Note, an E-value of 0.0 signifies a likelihood below 1e-250.

Why a Nobel Prize?

The real power of bioinformatics, especially BLAST, is twofold. First, bioinformatics methods dramatically amplify experimental results. Second, bioinformatics methods generate new testable hypotheses and targets for further discovery. To understand how BLAST amplifies biological knowledge, one must first have an appreciation of just how much in terms of time, energy, and resources it takes to make discoveries using the tools of molecular biology. Without basic information on the genetics and biochemical function of the genes or gene regions gleaned from laboratory experimentation, BLAST matches would not be particularly

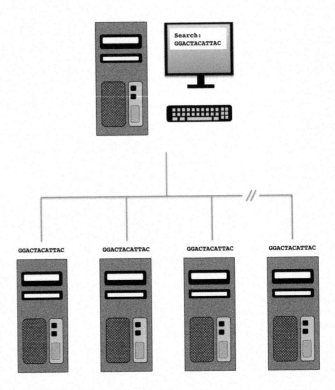

FIGURE 1.2. **Parallel BLAST searches.** Each "Query" sequence is simultaneously matched to hundreds or thousands of sequence databases using the BLAST algorithm. The scores for the best matches are retrieved and ranked.

useful. As an example, Fig. 1.3 shows some of the standard experimental approaches needed to determine the function of a protein-coding gene found in a bacterial genome.

While the cost of characterizing even a single protein-coding gene can be significant in terms of both time and money, once we have this characterization and this sequence in a database, algorithms like BLAST become especially powerful. Figure 1.3B illustrates how BLAST and DNA sequencing technology can amplify biological knowledge.

In the example, BLAST is really amplifying the experimental knowledge by allowing us to infer the function of 20 different previously unknown DNA sequences through a rapid sequence alignment. The process can also go in the other direction. For instance, we could use BLAST to discover if a DNA sequence that codes for a gene of unknown function is present in every living organism ever sequenced. However, should an experimentalist decide to target the protein product of this gene of unknown function for analysis and figure out its molecular function, we would suddenly know the function of this gene in millions of species.

Given the rate of sequence generation, it is safe to say that BLAST and other bioinformatics methods are the primary source of information for 99.9% of all newly sequenced DNA. This has made BLAST instrumental for discovering, among other things, the following:

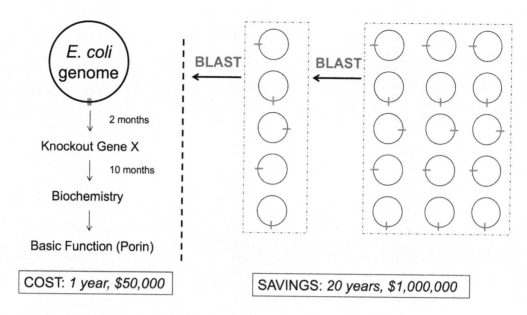

FIGURE 1.3. Characterizing the function of a novel bacterial protein encoded in the _Escherichia coli_ genome and then using BLAST to find similar functions in newly sequenced genomes. (Left) Flowchart of approaches necessary to characterize a novel protein in the _E. coli_ genome. Most bacteria have one copy of each gene on a single circular chromosome.[2] Each of the flowchart steps represents weeks or months of painstaking work. Typically, this process may take a year and cost on the order of $50,000. **(Right)** Once a gene has been characterized, sequences from other bacterial genomes (circles) can be quickly matched to this sequence using BLAST. The figure shows BLAST identifying porin proteins in 20 different bacterial genomes. When the matches are sufficiently strong, we can infer that these proteins have functions highly similar to that of the original, which saves the need to experimentally characterize this protein from all 20 genomes.

- Novel pathogens, both animal and human
- Life in extreme environments
- Mechanisms of genetic diseases
- Novel protein functions
- Novel cellular processes

If this doesn't warrant a Nobel Prize in Medicine, frankly, I don't know what does.

Notes

1. Note the FASTA format of the sequence, the program's enduring legacy to bioinformatics.
2. In the cell, the genomic DNA is usually tightly wound in a supercoiled state.

ACTIVITY 1.1   BLAST ALGORITHM

Motivation

The purpose of this activity is to teach the basic concepts behind the BLAST algorithm and how to use a web-based implementation of this algorithm to analyze DNA and protein sequence data. BLAST (Basic Local Alignment Search Tool) is a fast computational method for making sequence alignments. Sequence alignments are a critical part of bioinformatics. Computational methods for making pairwise alignments of biological molecules (DNA, RNA, or protein) were some of the very first bioinformatics algorithms developed. Among other things, sequence alignments allow researchers to determine the organisms from which the molecule came (human, oyster, pine tree, bacterium, etc.) and predict the cellular function of biological molecules based only on their sequence. For example, BLAST can report with high confidence that the protein sequence YNFGSGSAYGGSFGGVDGLLAGGEKATMQNL is keratin from the domestic dog hair found on your sofa.

BLAST was created to speed up the process of making sequence alignments. Full pairwise sequence alignment methods (see Chapter 03) are too computationally intensive to handle the alignment of thousands or millions of sequences. BLAST speeds up this process by "chopping up" an input sequence into smaller bits and matching these smaller bits to millions of different sequences. The algorithm then attempts to extend the sequence alignment to make a full alignment. Then the algorithm ranks the sequence alignments, and the longest alignment with the fewest mismatches wins! In bioinformatics parlance we call the good matches "hits," and the best ones are "best hits" or "top hits." BLAST is a heuristic method, meaning that it is not guaranteed to find the optimal alignment, but it is much faster than more stringent approaches.

Learning Objectives

1. Know the basic purpose and utility of the BLAST computational method (Motivation).
2. Understand the concepts behind the BLAST algorithm (Concepts and Exercises).
3. Correctly solve sequence-matching problems based on the BLAST algorithm (Concepts and Exercises).
4. Learn how to use NCBI's BLAST web-based sequence analysis website and be able to correctly interpret its output (Concepts and Exercises).

Concepts

As mentioned above, the purpose of the BLAST algorithm is to find the best hit (highest-scoring match) of an unknown DNA or protein sequence in a database. The method has been so successful because of its clever simplicity. In order to better grasp the method behind the algorithm, try the preparatory exercise. Using your brain and a pencil, try to find regions (local alignments) that best match the following DNA sequence and protein sequences. First, try to match the Query DNA sequence to the Sbjct[1] (Subject) DNA sequence. Then do the same for the protein Query and Sbjct sequences. Find the regions of best alignment between the two, and keep in mind that the match doesn't have to be perfect and may even need spaces to help it line up. Circle or draw lines between the matching letters.

DNA MATCH

Query: AGCGAATATTATGTTGAAGTAGCAAAGTCCTGGAGCCT

Sbjct: ACTACAGGGGAGTTTTGTTGAAGTTGCAAAGTCCTGGAGCCTCCAGAGGGC

PROTEIN MATCH

Query: MEMKATTALLNDRVLRAMLYFWCKAEETCALEVCEE

Sbjct: ETIRRAYPDANLLNDRVLRAMLYFWRKAEETCAPSVSMRKIVATWMLEVCEE

Reflection
* How much of the query did you try to match at one time?
* How did you find a match? Can you describe it in words?
* Were there any mismatches for the best sequence?
* Were there ever multiple matches? Would breaking up the Query (introducing a gap) help?

Below is the answer. The vertical lines indicate a perfect match between the letters of the two sequences. Note that there are some mismatches and that a big gap must be inserted for the end of the protein Query sequence to match the Sbjct sequence (LEVCEE).

DNA MATCH

Query: AGCGAATATTATGTTGAAGTAGCAAAGTCCTGGAGCCT
             || |||||||||| ||||||||||||||||||
Sbjct: ACTACAGGGGAGTTTTGTTGAAGTTGCAAAGTCCTGGAGCCTCCAGAGGGC

PROTEIN MATCH

Query: MEMKATTALLNDRVLRAMLYFWCKAEETCA———————————LEVCEE
            ||||||||||||| ||||||| ||||||
Sbjct: ETIRRAYPDANLLNDRVLRAMLYFWRKAEETCAPSVSMRKIVATWMLEVCEE

Most students solve this problem by (i) sliding the Query sequence along the Sbjct sequence, (ii) finding a short region that matches well, and then (iii) extending the match as far as possible. This is essentially how the BLAST algorithm works. Figure 1.1.1 details the basic steps of the algorithm.

Nucleotide BLAST | Protein BLAST

1) <u>Break the query sequence into words</u>
```
AGTACTAATAC
AGTA
 GTAC
  TACT
```
2) <u>Search for EXACT word matches</u>
AGTA
`CGCGAGCTGTAGCAAGTACTATTAC. . .`
3) <u>Extend the match until it falls below a fixed threshold or nothing left to match</u>
AGTACTAATAC
`CGCGAGCTGTAGCAAGTACTATTAC. . .`

1) <u>Break the query sequence into words</u>
```
RILYPAQVIGLR
RILY
 ILYP
  LYPA
```
2) <u>Search for EXACT word matches</u>
ILYP
`MVQGWALYDFLKCRAILYPGQVLMRWW...`
3) <u>Extend the match until it falls below a fixed threshold or nothing left to match</u>
RILYPAQVLR
`MVQGWALYDFLKCRAILYPGQVLMRWW...`

FIGURE 1.1.1. Principles behind the BLAST algorithm. (1) The first step of the algorithm is to break the Query sequence into smaller pieces called "words." For DNA, the word size is usually 10 or 11 letters long, but the example uses four-letter words for simplicity. (Four-letter words. *snicker*) Protein sequence matches start with fewer letters. (2) Then the algorithm slides these smaller words across possible target sequences until it finds a perfect match with one of the small words. (3) Starting with this small alignment, BLAST then extends the alignment until it runs out of letters or the alignment becomes poor (lots of mismatches). The score of the alignment is determined by summing up the scores for matches and mismatches. For DNA, BLAST uses +5 for a match and −4 for a mismatch. The scores of protein sequence alignments are determined by using a special table of match/mismatch scores, such as the BLOSUM62 matrix (see Chapter 07).

Exercises

Interactive exercise (theory)
Use the online BLAST Interactive Link below to learn how the algorithm makes and scores BLAST sequence alignments. Click on the dark circle with the yellow letter I at the top of the page to learn how to use the BLAST Exercise teaching interactive. Once you learn how it works, solve the activity problem.

<u>BLAST Interactive Link</u>
Link:
http://kelleybioinfo.org/algorithms /default.php?o=1

Problem

Practice with the BLAST Exercise Link, then solve the problem below.

1. Write the best alignment of the Query to each DNA sequence in the boxes and circle the first matching word from the Query.

2. Calculate scores and rank the three alignments.

Length		Scores	
Subject Size:	25	Match:	5
Query Size:	10	Mismatch:	-4
Word Size:	5		
Query:	GCACATGCCT		

Query: **G C A C A T G C C T**

DNA 1: **C A G A G A C C T C G A T C C T T A C A T G C C A**

DNA 2: **C A G A G A C C T C G A T G C A C A T G T C T T G**

DNA 3: **C A C A C A T G A C T T G G A C T C T G C C A T G**

	Score	Rank
DNA 1:		
DNA 2:		
DNA 3:		

Lab Exercises (Practice)

In this part of the exercise, you will learn how to analyze mystery DNA and protein sequences using the BLAST algorithm online at NCBI. You will also learn how to interpret the output from the program including what the values mean and how to find information about the best match in the database to your query sequence. You will also use a program, called ORF finder, that translates a DNA sequence into likely protein sequences.

NCBI BLAST Tutorial
Link:
**http://kelleybioinfo.org/algorithms
/tutorial/TAli1.pdf**

Sample and lab exercise data:
**http://kelleybioinfo.org/algorithms/data
/DAli1.txt**

Lab Exercise

Click on the sample and lab exercise data link for the sequence data used in this exercise.

Part 1

Use NCBI BLAST tools to analyze the following DNA that you just sequenced from a plasmid and answer the following questions:

>Part1_Plasmid_Derived_Sequence
CGTTTACGGCGTGGACTACCAGGGTATCTAATCCTGTTCGCTCCCCAACGCTTTCGCTCCTCAGCGT
CAGTTACTGCCCAGAGACCCGCCTTCGCCACCGGTGTTCCTCCTGATATCTGCGCATTCCACCGCTA
CACCAGGAATTCCAGTCTCCCCTGC

1. Use the NCBI BLAST tool to perform a sequence search with the above sequence.

    a. The highest-scoring BLAST hit is to what named organism? (Ignore the unknown/uncultured organism hits.)

    b. What is the gene name?

    c. What is the function of the gene, if known? (Don't know? Try asking "Professor" Google or "Dr." Wikipedia!)

    d. Who submitted the sequence?

    e. From what institution?

    f. Get the following data for this particular match.

        i. E-value:

        ii. Identities:

Part 2

>Part2_Protein_Sequence
MNGTEGPNFYVPFSNKTGVVRSPFEYPQYYLAEPWQFSMLAAYMFLLIVLGFPINFLTLYVTVQHKK
LRTPLNYILLNLAVANHFMVFGGFTTTLYTSLHGYFVFGSTGCNLEGFFATLGGEIALWSLVVLAIE
RYVVVCKPMSNFRFGENHAIMGVAFTWVMALACAAPPLVGWSRYIPEGMQCSCGIDYYTLKPEVNNE
SFVIYMFVVHFTIPMTIIFFCYGQLVFTVKEAAAQQQESATTQKAEKEVTRMVIIMVIAFLICWVPY
ASVAFYIFTHQGSDFGPILMTLPAFFAKSSAIYNPVIYIMMNKQFRNCMLTTICCGKNPFGEEEGST
TASKTETSQVAPA

1. Use the NCBI BLAST tool to perform a sequence search with the protein sequence above.

   a. The highest-scoring BLAST hit is to what organism?

   b. What is the gene?

   c. What is the function of the gene, if known?

   d. Who submitted the sequence?

   e. From what institution?

   f. Get the following data for this particular match.

      i. E-value:

      ii. Identities:

2. Use the NCBI ORF finder (**https://www.ncbi.nlm.nih.gov/orffinder/**) to translate the Mystery DNA below. This sequence came from a bacterial culture, so use the bacterial genetic code.

   a. First of all, what exactly is an "ORF" anyway?

   b. What are the nucleotide positions of the longest ORF?

   c. How many different reading frames did the program reveal that were longer than 100 amino acids (aa)?

   d. BLAST the longest putative ORF.

      i. What is the name of the gene?

      ii. What is the name of the organism?

>Part2_Mystery_DNA
TCCCTCCACAAAGAATGGAGCTGTGAACTACTAGCACGCAATGTGATTCCTGCAATTGAAAATGAACAAT
ATATGCTACTTATAGATAACGGTATTCCGATCGCTTATTGTAGTTGGGCAGATTTAAACCTTGAGACTGA
GGTGAAATATATTAAGGATATTAATTCGTTAACACCAGAAGAATGGCAGTCTGGTGACAGACGCTGGATT
ATTGATTGGGTAGCACCATTCGGACATTCTCAATTACTTTATAAAAAAATGTGTCAGAAATACCCTGATA
TGATCGTCAGATCTATACGCTTTTATCCAAAGCAGAAAGAATTAGGCAAAATTGCCTACTTTAAAGGAGG
TAAATTAGATAAAAAAACAGCAAAAAAACGTTTTGATACATATCAAGAAGAGCTGGCAACAGCACTTAAA
AATGAATTTAATTTTATTAAAAAAATAGAAGGAGACATCCCTTATGGGAACTAGACTTACAACCCTATCAA
ATGGGCTAAAAAACACTTTAACGGCAACCAAAAGTGGCTTACATAAAGCCGGTCAATCATTAACCCAAGC
CGGCAGTTCTTTAAAAACTGGGGCAAAAAAAATTATCCTCTATATTCCCCAAAATTACCAATATGATACT
GAACAAGGTAATGGTTTACAGGATTTAGTCAAAGCGGCCGAAGAGTTGGGGATTGAGGTACAAAGAGAAG
AACGCAATAATATTGCAACAGCTCAAACCAGTTTAGGCACGATTCAAACCGCTATTGGCTTAACTGAGCG
TGGCATTGTGTTATCCGCTCCACAAATTGATAAATTGCTACAGAAAACTAAAGCAGGCCAAGCATTAGGT
TCTGCCGAAAGCATTGTACAAAATGCAAATAAAGCCAAAACTGTATTATCTGGCATTCAATCTATTTTAG
GCTCAGTATTGGCTGGAATGGATTTAGATGAGGCCTTACAGAATAACAGCAACCAACATGCTCTTGCTAA
AGCTGGCTTGGAGCTAACAAATTCATTAATTGAAAATATTGCTAATTCAGTAAAAACACTTGACGAATTT
GGTGAGCAAATTAGTCAATTTGGTTCAAAACTACAAAATATCAAAGGCTTAGGGACTTTAGGAGACAAAC
TCAAAAATATCGGTGGACTTGATAAAGCTGGCCTTGGTTTAGATGTTATCTCAGGGCTATTATCGGGCGC

```
AACAGCTGCACTTGTACTTGCAGATAAAAATGCTTCAACAGCTAAAAAAGTGGGTGCGGGTTTTGAATTG
GCAAACCAAGTTGTTGGTAATATTACCAAAGCCGTTTCTTCTTACATTTTAGCCCAACGTGTTGCAGCAG
GTTTATCTTCAACTGGGCCTGTGGCTGCTTTAATTGCTTCTACTGTTTCTCTTGCGATTAGCCCATTAGC
ATTTGCCGGTATTGCCGATAAATTTAATCATGCAAAAAGTTTAGAGAGTTATGCCGAACGCTTTAAAAAA
TTAGGCTATGACGGAGATAATTTATTAGCAGAATATCAGCGGGGAACAGGGACTATTGATGCATCGGTTA
CTGCAATTAATACCGCATTGGCCGCTATTGCTGGTGGTGTGTCTGCTGCTGCAGCCGGCTCGGTTATTGC
TTCACCGATTGCCTTATTAGTATCTGGGATTACCGGTGTAATTTCTACGATTCTGCAATATTCTAAACAA
GCAATGTTTGAGCACGTTGCAAATAAAATTCATAACAAAATTGTAGAATGGGAAAAAAATAATCACGGTA
AGAACTACTTTGAAAATGGTTACGATGCCCGTTATCTTGCGAATTTACAAGATAATATGAAATTCTTACT
GAACTTAAACAAAGAGTTACAGGCAGAACGTGTCATCGCTATTACTCAGCAGCAATGGGATAACAACATT
GGTGATTTAGCTGGTATTAGCCGTTTAGGTGAAAAAGTCCTTAGTGGTAAAGCCTATGTGGATGCGTTTG
AAGAAGGCAAACACATTAAAGCCGATAAATTAGTACAGTTGGATTCGGCAAACGGTATTATTGATGTGAG
TAATTCGGGTAAAGCGAAAACTCAGCATATCTTATTCAGAACGCCATTATTGACGCCGGGAACAGAGCAT
CGTGAACGCGTACAAACAGGTAAATATGAATATATTACCAAGCTCAATATTAACCGTGTAGATAGCTGGA
AAATTACAGATGGTGCAGCAAGTTCTACCTTTGATTTAACTAACGTTGTTCAGCGTATTGGTATTGAATT
AGACAATGCTGGAAATGTAACTAAAACCAAAGAAACAAAAATTATTGCCAAACTTGGTGAAGGTGATGAC
AACGTATTTGTTGGTTCTGGTACGACGGAAATTGATGGCGGTGAAGGTTACGACCGAGTTCACTATAGCC
GTGGAAACTATGGTGCTTTAACTATTGATGCAACCAAAGAGACCGAGCAAGGTAGTTATACCGTAAATCG
TTTCGTAGAAACCGGTAAAGCACTACACGAAGTGACTTCAACCCATACCGCATTAGTGGGCAACCGTGAAG
AAAAAATAGAATATCGTCATAGCAATAACCAGCACCATGCCGGTTATTACACCAAAGATACCTTGAAAG
CTGTTGAAGAAATTATCGGTACATCACATAACGATATCTTTAAAGGTAGTAAGTTCAATGATGCCTTTAA
CGGTGGTGATGGTGTCGATACTATTGACGGTAACGACGGCAATGACCGCTTATTTGGTGGTAAAGGCGAT
GATATTCTCGATGGTGGAAATGGTGATGATTTTATCGATGGCGGTAAAGGCAACGACCTATTACACGGTG
GCAAGGGCGATGATATTTTCGTTCACCGTAAAGGCGATGGTAATGATATTATTACCGATTCTGACGGCAA
TGATAAATTATCATTCTCTGATTCGAACTTAAAAGATTTAACATTTGAAAAAGTTAAACATAATCTTGTC
ATCACGAATAGCAAAAAAGAGAAAGTGACCATTCAAAACTGGTTCCGAGAGGCTGATTTTGCTAAAGAAG
TGCCTAATTATAAAGCAACTAAAGATGAGAAAATCGAAGAAATCATCGGTCAAAATGGCGAGCGGATCAC
CTCAAAGCAAGTTGATGATCTTATCGCAAAAGGTAACGGCAAAATTACCCAAGATGAGCTATCAAAAGTT
GTTGATAACTATGAATTGCTCAAACATAGCAAAAATGTGACAAACAGCTTAGATAAGTTAATCTCATCTG
TAAGTGCATTTACCTCGTCTAATGATTCGAGAAATGTATTAGTGGCTCCAACTTCAATGTTGGATCAAAG
TTTATCTTCTCTTCAATTTGCTAGAGCAGCTTAATTTTTAATGATTGGCAACTCTATATTGTTTCACACA
TTATAGAGTTGCCGTTTTATTTTATAAAAGGAGACAATATGGAAGCTAACCATCAAAGGAATGATCTTGG
TTTAGTTGCCCTCACTATGTTGGCACAATACCATAATATTTCGCTTAATCCGGAAGAAATAAAACATAAA
```

PROTEIN ANALYSIS

P roteins are the workhorses of biology. There is a reason the central dogma of molecular biology ends in the formation of proteins: every cellular and physiological process within an organism involves proteins. Whether it is for synthesizing RNA or DNA, propagating signals along nerve fibers, or destroying disease-causing pathogens, proteins (mainly enzymes) do most or all of the work. Proteins also determine the shape and mobility of cells (structural proteins), act as gatekeepers for molecules entering and exiting cell membranes (membrane proteins), respond to external signals such as hormones or glucose (receptor proteins), and transmit signals inside of cells (cell signaling proteins).

Figure 2.1 presents a few examples of proteins that perform common cellular processes.

Protein Bioinformatics

In order to understand the biological function of a protein, one must first understand its physical properties. Ideally, one would know the entire three-dimensional (3D) structure of a protein and the location and angle of every amino acid in the protein at the atomic level. Unfortunately, experimental determination of protein 3D structure, primarily performed by first crystallizing the protein, is often a monumental undertaking that is not possible for all types of structures (e.g., proteins inside cell membranes known as transmembrane proteins). Protein crystal structures can take months or years to determine, and many labs devote their entire research program to the crystallization of important proteins.

Naturally, slow and methodical experimental methods cannot possibly keep up with the rate at which scientists are generating protein sequence data. Sequencing technologies can generate the entire sequence of multiple bacterial genomes in a day. Given that an average sized bacterial genome may encode >3,000 protein sequences, you can clearly see the need for computational methods to speed up this process of determining protein functions. To fill this need,[1] bioinformaticians have designed algorithms to leverage our vast existing knowledge of amino acids and protein structure to develop algorithms that can determine structural or functional properties of proteins using only the primary protein sequence. This is

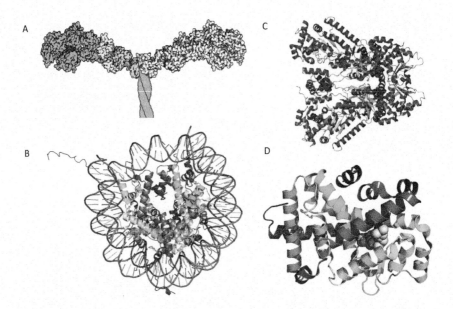

FIGURE 2.1. Examples of proteins involved in important organismal or cellular processes. (A) Myosin. Image courtesy of David S. Goodsell/RCSB PDB, under license CC BY-4.0. **(B)** Complex between nucleosome core particle and DNA fragment. Image courtesy of Emw, based on PDB ID 1aoi, under license CC BY-3.0. **(C)** Cartoon representation of anthrax toxin. Image courtesy of the European Bioinformatics Institute (**http://www.ebi.ac.uk/**). **(D)** Crystal structure of ligand binding domain of RORγ and ligand.

great because it is really easy to determine the primary sequence of a protein given its DNA sequence: simply run the DNA sequence through a genetic code translator and voilà—you have the protein sequence! Figure 2.2 illustrates the basic process of computational translation of a DNA sequence that contains a protein-coding sequence.

Bioinformatics Methods

The algorithms we cover in this chapter attempt to predict physical and structural properties of a protein using only its primary sequence, literally a text string of amino acid letters. (Like this protein sequence of unknown function: DRKELLEYISRAD.) Of course, the fastest methods of determining the structure of a novel unknown protein sequence is to find a highly similar match in a database to a protein with known structure. For example, it would be easy to determine most of the structure and function of a newly sequenced bacterial outer membrane protein (OMP) because there are many well-characterized OMPs with 3D structures already in GenBank, UniProt, and other databases. Matches to more distantly related sequences with similar structures can also reveal important structural features. The Pfam (protein family) database, COG (clusters of orthologous groups) database, and other databases use various algorithms, such as multiple-sequence alignments and hidden Markov models, to group proteins into various functional classes. A significant match to a family or cluster could provide insight into the protein's structure or function.

Original DNA sequence

AATGATGAAGCGCAATATTCTGGCAGTGATCGTCCCTGCTCTGTTAGTAGCAGGTACTGCA

↓

6 Frame Protein Translation

5'3' Frame 1
N D E A Q Y S G S D R P C S V S S R Y C

5'3' Frame 2
Met Met K R N I L A V I V P A L L V A G T A

5'3' Frame 3
Stop Stop S A I F W Q Stop S S L L C Stop Stop Q V L

3'5' Frame 1
C S T C Y Stop Q S R D D H C Q N I A L H H

3'5' Frame 2
A V P A T N R A G T I T A R I L R F I I

3'5' Frame 3
Q Y L L L T E Q G R S L P E Y C A S S

FIGURE 2.2. **Computational translation.** Once the DNA sequence is determined, we can use the genetic code to do a translation with the computer. We can skip over the transcription step here, because the RNA is essentially a copy of the DNA sequence. Since we initially do not know which strand of the DNA sequence is read by the RNA polymerase, it is best to translate both the DNA sequence and its reverse complement (the other strand). We also do not necessarily know the frame (reading frame). The ribosome reads the RNA in groups of nucleotide triplets, called codons.[2] By shifting the starting point of the ribosome, you change the reading frame for the entire sequence. In this figure, frame 1 corresponds to a ribosome starting on the first A of the original DNA sequence, frame 2 corresponds to a ribosome starting on the second A, and frame 3 corresponds to a ribosome starting on the first T. When determining a potential protein sequence from a given DNA sequence, unless one knows the starting codon, the best thing is to translate all 6 possible reading frames and pick the longest one that doesn't have a stop codon. In practice, one can also BLAST all 6 possible protein sequences and pick the one that matches a known protein in the database.

The problem becomes more difficult as the protein becomes more dissimilar to known proteins. In this chapter, we cover methods that predict the properties of protein sequences based only on the amino acid composition of the sequence. The first method we discuss predicts the relative hydrophobicity of protein sequences by scanning the sequence and looking for regions of the protein with lots of hydrophobic amino acids. Hydrophobic means water (hydro-) fearing (-phobic). Protein regions that are hydrophobic are usually found interacting with hydrophobic molecules, such as the lipids found in the cell membrane. Figure 2.3 illustrates some examples of proteins with critical hydrophobic regions.

The second method we cover is a probability approach to determine the secondary structure of a novel protein using the primary sequence. Protein 3D structures

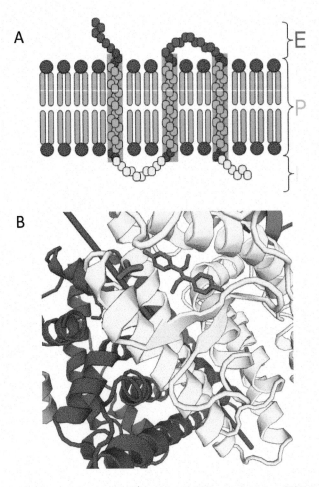

FIGURE 2.3. Examples of hydrophobic regions in different proteins. (A) A hypothetical transmembrane protein (red, green, and yellow ovals) passes through a cell membrane lipid bilayer (blue circles, hydrophilic lipid bilayer heads; green oblong shapes, hydrophobic tails). E, extracellular (outside the cell); I, intracellular (inside the cell); P, plasma membrane (also just the cell membrane). The hydrophobic regions of the protein (green ovals) help the protein localize in the cell membrane. Image courtesy of Magnus Manske, under license CC BY-3.0. **(B)** Close-up of a steroid hormone-binding transcription factor (human estrogen receptor). The steroid hormone (molecule with symmetrical rings in blue, center top) binds in the hydrophobic pocket of the protein, changing the 3D structure of the receptor so it can activate gene expression.[3] Steroid hormones are lipids (fats), which are hydrophobic molecules. This is why the binding pocket also needs to be hydrophobic or it would repel the steroid hormone molecules.

are really combinations of two main secondary structural elements, alpha helices and beta sheets, with loop regions connecting the various structural elements. Figure 0.8 illustrates how the combinations of protein secondary structure elements can combine to form the final 3D structure of a protein. The Chou-Fasman algorithm that we look at in this chapter uses knowledge gleaned from many protein sequences to determine how often particular amino acids are found in alpha helices or beta sheets.

Notes

1. "I feel the need, the need for speed."—Tom Cruise, pilot and bioinformatician.
2. The ribosome matches the codon with the tRNA anti-codon to covalently bind together the amino acids during protein synthesis.
3. An example of this is the androgen receptor, which binds the steroid hormone testosterone and acts as a transcription factor to alter the expression of muscle-building genes, especially in "roid ragers." Or professional athletes.

ACTIVITY 2.1   HYDROPHOBICITY PLOTTING

Motivation

The purpose of this activity is to teach the concepts behind hydrophobicity plotting. Hydrophobicity plots use a simple sequence scanning approach and experimental values of amino acid hydrophobicity to determine which parts of a protein are hydrophobic. Generally speaking, hydrophobicity is one of the most important properties of biological molecules. Proteins tend to be a mix of both hydrophobic and hydrophilic amino acids. Hydrophobic ("water-fearing") proteins tend to be found in parts of the cell that are rich in lipids, such as the cell membrane, the nuclear membrane, and vesicles. If we were to determine that a protein had many long sections of hydrophobic amino acids, this could mean that the protein was part of a membrane or vesicle. Similarly, if we were to determine that a part of the protein was hydrophilic ("water loving"), we could predict that this part of the protein was not likely to be found in a membrane or binding lipid molecules. Instead, we might predict that it was intra- or extracellular or bound to hydrophilic charged molecules like DNA or RNA. In this section, you will learn a basic method of hydrophobicity plotting for predicting hydrophobic (or hydrophilic) regions of any given protein sequence. Afterwards, you will gain practice with online software that determines the most hydrophobic regions of proteins using the same algorithm.

Learning Objectives

1. Understand the basic concept of protein hydrophobicity and why it is important for understanding protein function (Motivation).
2. Learn how to use a sliding window algorithm and an amino acid hydrophobicity scale to determine the most hydrophobic region of a protein (Concepts and Exercises).
3. Use online hydrophobicity software to plot hydrophobic regions of a protein sequence and interpret the output (Concepts and Exercises).

Concepts

To better understand the method behind hydrophobicity plotting, try the preparatory exercise on the next page, which asks you to circle or underline the most hydrophobic region of the protein sequence "Protein Seq 1" using information in the diagram. The diagram indicates the amount of free energy needed to dissolve each of the amino acids found in biological protein sequences into a hydrophobic solvent. The more negative the change in free energy value (ΔG), the more readily the amino acid dissolves into the solvent (in this case, octanol) and the more hydrophobic the amino acid. For example, leucine (L) has a ΔG of −1.2. The more hydrophilic amino acids have high positive ΔG values (e.g., histidine [H] has a ΔG of +2.4).

Protein Seq 1:

E R K P Y L V A W M K R

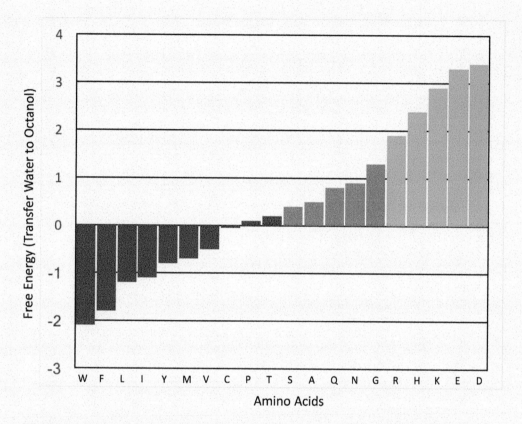

FIGURE 2.1.1. Free energies for transfer of amino acids from water to octanol. Green bars, charged residues; orange bars, polar residues; purple bars, hydrophobic residues. Data from **Bowie JU**. 2005. *Nature* **438:**581–589.

Reflection

- How did you find the beginning of the hydrophobic region?
- Was searching for the hydrophobic region similar in any way to the BLAST algorithm? Why or why not?
- Could you use the free energy diagram (Fig. 2.1.1) to score this hydrophobic region?
- Plot the average hydrophobicity scores of the first 4 amino acids, the middle 4 amino acids, and the final 4 amino acids in the box on the next page.

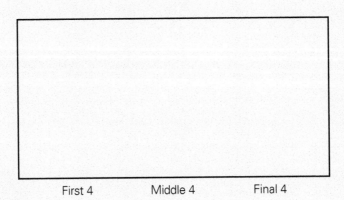

First 4   Middle 4   Final 4

The most hydrophobic region of the sequence is underlined below. To determine a score for this region, one might sum up the total free energy score of this 6-amino-acid-long region. Better yet, the total could be divided by the length of the hydrophobic region to determine the average hydrophobicity of this region. In this case, the most hydrophobic region would be negative.

Protein Seq 1:

E R K P Y L V A W M K R

The hydrophobicity plotting method you will learn using the interactive module also scans protein sequences a few amino acids at a time and calculates the average hydrophobicity. However, instead of free energy values, we will use the hydropathy scores developed by Kyte and Doolittle (a Kyte-Doolittle hydropathy plot), in which the most hydrophobic amino acids have large positive values. For example, the Kyte-Doolittle hydropathy value for the hydrophobic amino acid leucine is +3.8, while the hydropathy of the polar amino acid histidine is −3.2.

Exercises

Interactive exercise (theory)
Use the online hydrophobicity exercise link below to learn how to find the hydrophobic regions of a protein sequence. The Interactive Link explains how to use the teaching interactive. Once you learn how it works, solve the activity problem.

Hydrophobicity Interactive Link
Link:
**http://kelleybioinfo.org/algorithms
/default.php?o=2**

Problem

1. Fill in the two empty boxes using the amino acid hydrophobicity scores below.

2. Draw the hydrophobicity plot in the gray box below.

3. Write the score and average for the last 5-amino-acid window.

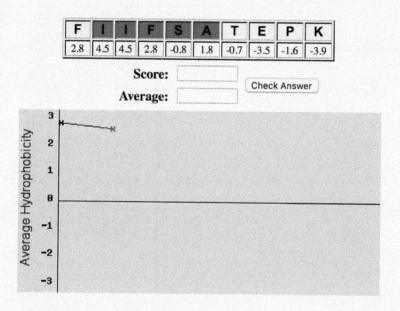

F		I	F	S	A	T	E	P	K
2.8	4.5	4.5	2.8	-0.8	1.8	-0.7	-3.5	-1.6	-3.9

Score: [   ]

Average: [   ]

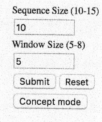
Check Answer

Amino Acid	H-Score	Amino Acid	H-Score
A	1.8	L	3.8
R	-4.5	K	-3.9
N	-3.5	M	1.9
D	-3.5	F	2.8
C	2.5	P	-1.6
Q	-3.5	S	-0.8
E	-3.5	T	-0.7
G	-0.4	W	-0.9
H	-3.2	Y	-1.3
I	4.5	V	4.2

Sequence Size (10-15)

[10]

Window Size (5-8)

[5]

[Submit] [Reset]

[Concept mode]

Lab Exercises (Practice)

In this part of the exercise, you will learn how to determine the hydrophobic regions of protein sequences using the ProtScale website. You will also use a program called TMHMM, which uses hidden Markov models to find transmembrane proteins.

ProtScale Tutorial Link
Link:
**http://kelleybioinfo.org/algorithms
/tutorial/TPro1.pdf**

Sample and lab exercise data:
**http://kelleybioinfo.org/algorithms
/data/DPro1.txt**

Lab Exercise

>ProteinSequence21A

**MGPTSVPLVKAHRSSVSDYVNYDIIVRHYNYTGKLNISADKENSIKLTSVVFILICCFIILENIFVL
LTIWKTKKFHRPMYYFIGNLALSDLLAGVAYTANLLLSGATTYKLTPAQWFLREGSMFVALSASVFS
LLAIAIERYITMLKMKLHNGSNNFRLFLLISACWVISLILGGLPIMGWNCISALSSCSTVLPLYHKH
YILFCTTVFTLLLLSIVILYCRIYSLVRTRSRRLTFRKNISKASRSSEKSLALLKTVIIVLSVFIAC
WAPLFILLLLDVGCKVKTCDILFRAEYFLVLAVLNSGTNPIIYTLTNKEMRRAFIRIMSCCKCPSGD
SAGKFKRPIIAGMEFSRSKSDNSSHPQKDEGDNPETIMSSGNVNSSS**

1. Determine the following about ProteinSequence21A:

   a. What is this protein? (Hint: you may need to use another algorithm that we have already covered to search for a close match to this sequence.)

   b. Create a hydrophobicity plot (window size = 19) with this sequence data using the Kyte-Doolittle hydrophobicity scale. Draw/show the plot below.

   c. Repeat the plotting using the Eisenberg hydrophobicity scale, a scale based on different free energy values. Describe briefly how they compare in terms of transmembrane predictions?

2. Use ProteinSequence21A with TMHMM. Draw/show the TMHMM plot below.

   a. How do the plot results compare to the Kyte-Doolittle and Eisenberg graph predictions? Explain.

ACTIVITY 2.2   PROTEIN SECONDARY STRUCTURE PREDICTION

Motivation

As described in the Chapter 02 introduction, the structure of a protein can be described at three different levels: primary (also written as 1°), secondary (2°), and tertiary (3°). The primary structure is the linear order of amino acids, and the tertiary structure is the full 3D structure. In between primary and tertiary structures stands the secondary structure, which also tells a great deal about protein structure and is easier to predict than the tertiary structure. The two basic types of secondary structures are alpha helices and beta sheets. There are also loop regions, which mainly serve to connect the helical and sheet structural elements. Proteins are essentially alpha helices, beta sheets, and loops arranged in different combinations. Some tertiary structures are formed exclusively from one type of secondary structure (e.g., the structure known as a beta barrel), but more often they have a mix of alpha helices and beta sheets.

The purpose of this activity is to teach the Chou-Fasman algorithm for predicting protein secondary structure based on the primary protein sequence. The Chou-Fasman algorithm was first designed in the early 1970s by, you guessed it, Chou and Fasman. By looking at amino acids in proteins with known structure, Chou and Fasman determined how often each amino acid appeared in an alpha helix, beta sheet, or loop region. They then assigned a likelihood score, called a propensity, for each of the 20 most common amino acids. Some amino acids are more common in alpha helices than in beta sheets and vice versa. In this chapter, you will learn how to use the propensities to score a protein sequence as being an alpha-helical region or a beta sheet and then use online prediction software on actual protein sequences.

Learning Objectives

1. Understand the basic concept of protein secondary structure, the two main forms (alpha helix and beta sheet), and why it is important for understanding protein function (Motivation).
2. Learn the principles behind the Chou-Fasman algorithm and be able to calculate the relative scores for the likelihood of a protein sequence forming an alpha helix or a beta sheet (Concepts and Exercises).
3. Learn how to use a sliding-window algorithm and secondary structure propensity scores to predict whether a section of a protein is an alpha helix or a beta sheet (Concepts and Exercises).
4. Use online secondary structure prediction (Chou-Fasman) software to predict protein secondary sequence and interpret the output (Concepts and Exercises).

Concepts

To understand how to use the Chou-Fasman algorithm, one must first understand the principle of propensity and how propensity values are calculated. The word propensity means an inclination or natural tendency to behave in a particular way. For example, I have a propensity for nerdiness. (Set phasers to stun!) The secondary structure propensity of an amino acid is equivalent to the probability (or the likelihood) of an amino acid being part of a particular type of secondary structure. Three propensities are calculated for each amino acid:

P(a) = the propensity of it being in an alpha helix
P(b) = the propensity of it being in a beta sheet
P(turn) = the propensity of it being in a turn

Propensities greater than 100 mean that an amino acid is more likely than by random chance of being in that particular structure, while propensities less than 100 mean that the amino acid is less likely to be found in that type of secondary structure.

Here are three examples:

Amino Acid	P(a)	P(b)	P(turn)
Alanine (A)	**142**	83	66
Threonine (T)	83	**119**	96
Asparagine (R)	67	89	**156**

These propensity values indicate that alanines are more likely to be found in alpha helices than chance would dictate and less likely to be in beta sheets or turns. Threonines have a higher propensity to be in beta sheets than in alpha helices and turns, and asparagines are more likely to be in turns than in alpha helices or beta sheets. To be clear, just because alanines have a higher propensity to be in alpha helices does not mean that they cannot also be in beta sheets or turns, but it does mean that if you find a region of a protein sequence with more alanines, this region is more likely to be an alpha helix.

Calculating Propensities

How are propensities calculated? Well, like most great bioinformatics values, propensities are based on real (experimental) data. In this case, amino acid secondary structure propensities are based on how many times each amino acid is found in alpha helices, beta sheets, and turns in known protein structures. Figure 2.2.1 shows an example of how to calculate the alpha helix propensity [P(a)] of lysines.

Reflection 1
- Based on the data in Figure 2.2.1, what is the propensity for lysines being in a beta sheet? A turn?
- If the expected value for lysine being in a turn was 0.1 because only 10% (0.1) of the amino acids in the database were in turns, how would this change the P(turn) for lysine?
- Could you use these numbers to determine whether a region of a protein was likely to be an alpha helix?

Where's Alphie?

Like "Where's Waldo" but not as frustrating!
The table on page 61 shows the propensities of each amino acid for being in an alpha helix, beta sheet, or turn. Use the values to determine which region of the Protein Seq 2 might be an alpha helix.

Protein Seq 2:

A E E M H L R N G I Q C Q W Y F

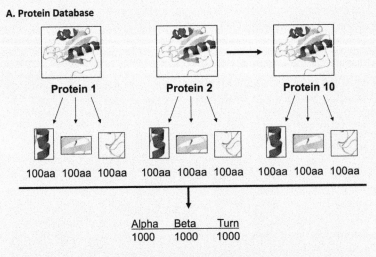

A. Protein Database

Protein 1   Protein 2   Protein 10

100aa 100aa 100aa   100aa 100aa 100aa   100aa 100aa 100aa

Alpha	Beta	Turn
1000	1000	1000

3000 total amino acids

B. Propensity calculation.

(1) Calculate the expected proportion of a given amino acid being in an alpha helix, beta sheet or a turn.

Alpha	Beta	Turn		
1000	1000	1000	=	3000 total amino acids
.33	.33	.33		proportion in each type

(2) How many total Lysine are in the database? How many are in each type of secondary structure?

Lysines:

Total number of Lysines in all 10 proteins	= 100
Number of Lysines in alpha helices	= 80
Number of Lysines in beta sheets	= 15
Number of Lysines in Turns	= 5

(3) The Observed and Expected proportions of Lysines in an Alpha Helix.

Observed (in helices)	= .80
Expect (at random)	= .33

(4) The propensity is just the Observed divided by the Expected!

Propensity = .80/.33   =   2.42 (*100 = 242)

FIGURE 2.2.1. **Calculation of P(a) for lysines. (A)** A protein database with 10 proteins, each with 300 amino acids evenly split among alpha helices, beta sheets, and turns. **(B)** To calculate the P(a), first determine the underlying probability of any given amino acid being in an alpha helix. From the database we know that it is 1/3 (0.33), since 1/3 of all the amino acids in the database have been experimentally determined to be in alpha helices. Focusing just on lysines, of the 100 lysines found in the database, most of them are in alpha helices (0.8 or 80%), which suggests a propensity of lysines to be in alpha helices. The observed P(a) for lysines is 0.80 (80/100), while the expected value is 0.33 for any amino acid to be in an alpha helix. The propensity is pretty easy to calculate: divide the observed value by the expected value and multiply by 100.

Amino Acid	P(a)	P(b)	P(turn)
Alanine (A)	142	83	66
Arginine (R)	98	93	95
Asparagine (N)	67	89	156
Aspartic acid (D)	101	54	146
Cysteine (C)	70	119	119
Glutamic acid (E)	151	37	74
Glutamine (Q)	111	110	98
Glycine (G)	57	75	156
Histidine (H)	100	57	95
Isoleucine (I)	108	160	47
Leucine (L)	121	130	59
Lysine (K)	114	74	101
Methionine (M)	145	105	60
Phenylalanine (F)	113	138	60
Proline (P)	57	55	152
Serine (S)	77	75	143
Threonine (T)	83	119	96
Tryptophan (W)	108	137	96
Tyrosine (Y)	69	147	114
Valine (V)	106	170	50

Reflection 2

- How does searching for alpha helices compare to hydrophobicity plotting or BLAST?
- How many amino acids have a high propensity for being in both alpha helices and beta sheets?
- Are there any regions in the sequence that might be more likely to be beta sheets?

Below is the answer. The underlined region has a large proportion of amino acids that tend to be found in alpha helices. The middle region is more likely to be a turn (italics), and the right side is more likely to be a beta sheet (blue).

<u>A E E M H L</u> *R N G* I Q C Q W Y *F*

Exercises

Interactive exercise (theory)

Use the online exercise link below to learn how to predict the secondary structure of a protein sequence. The Interactive Link explains how to use the teaching interactive. Once you learn how it works, solve the activity problem.

Chou-Fasman Interactive Link
Link:
http://kelleybioinfo.org/algorithms /default.php?o=9

Problem

1. Circle the "start" and "stop" regions of the protein sequence using the Chou-Fasman algorithm for an alpha helix.

2. Calculate the score for the alpha helix.

3. Is it an alpha helix or not? Answer yes or no below and indicate why you reached this conclusion. (Hint: calculate the score for the exact same region of the sequence being a beta sheet. Which is greater, the alpha helix or the beta sheet score?)

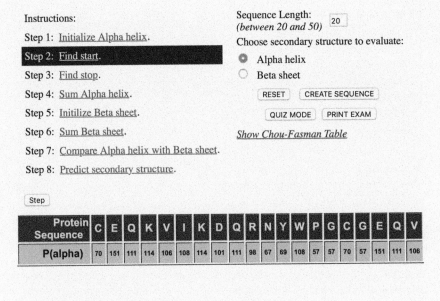

Instructions:

Step 1: Initialize Alpha helix.

Step 2: Find start.

Step 3: Find stop.

Step 4: Sum Alpha helix.

Step 5: Initilize Beta sheet.

Step 6: Sum Beta sheet.

Step 7: Compare Alpha helix with Beta sheet.

Step 8: Predict secondary structure.

Sequence Length:
(between 20 and 50)   20

Choose secondary structure to evaluate:

⦿ Alpha helix
◯ Beta sheet

RESET   CREATE SEQUENCE

QUIZ MODE   PRINT EXAM

Show Chou-Fasman Table

Step

Protein Sequence	C	E	Q	K	V	I	K	D	Q	R	N	Y	W	P	G	C	G	E	Q	V
P(alpha)	70	151	111	114	106	108	114	101	111	98	67	69	108	57	57	70	57	151	111	106

SCORE FOR ALPHA HELIX: _____
IS IT AN ALPHA HELIX? ___ WHY or WHY NOT?

Lab Exercises (Practice)

In this part of the exercise, you will learn how to analyze protein sequences using the Chou-Fasman algorithm online. You will also learn how to interpret the output from the program, including what the values mean and how to find information about the best match in the database to your query sequence.

You will also use the UniProt website and the ProtParam tool to analyze protein sequences.

ProtScale Tutorial Link
Link:
**http://kelleybioinfo.org/algorithms/tutorial
/TPro1.pdf**

Sample and lab exercise data:
**http://kelleybioinfo.org/algorithms
/data/DPro1.txt**

Lab Exercise

>ProteinSequence22A

MWVLINLLILMIMVLISVAFLTLLERKILGYIQDRKGPNKIMLFGMFQPFSDALKLLSKEWFFFNYSNLFIYSPMLMFFLS
LVMWILYPWFGFMYYIEFSILFMLLVLGLSVYPVLFVGWISNCNYAILGSMRLVSTMISFEINLFFLVFSLMMMVESFSFN
EFFFFQNNIKFAILLYPLYLMMFTSMLIELNRTPFDLIEGESELVSGFNIEYHSSMFVLIFLSEYMNIMFMSVILSLMFYG
FKYWSIKFILIYLFHICLIIWIRGILPRIRYDKLMNMCWTEMLMLVMIYLMYLYFMKEFLCI

1. Answer the following questions about ProteinSequence22A.

    a. First of all, what is this protein? (BLAST on UniProt website **http://www
        .uniprot.org/blast/**)

    Use the ProtParam tool (**http://web.expasy.org/protparam/**) for the next
    questions.

    b. What is the molecular weight?

    c. What is the number of positively charged residues?

>ProteinSequence22B

MGPTSVPLVKAHRSSVSDYVNYDIIVRHYNYTGKLNISADKENSIKLTSVVFILICCFIILENIFVLLTIWKTKKFHRPMY
YFIGNLALSDLLAGVAYTANLLLSGATTYKLTPAQWFLREGSMFVALSASVFSLLAIAIERYITMLKMKLHNGSNNFRLFL
LISACWVISLILGGLPIMGWNCISALSSCSTVLPLYHKHYILFCTTVFTLLLLSIVILYCRIYSLVRTRSRRLTFRKNISK
ASRSSEKSLALLKTVIIVLSVFIACWAPLFILLLLDVGCKVKTCDILFRAEYFLVLAVLNSGTNPIIYTLTNKEMRRAFIR
IMSCCKCPSGDSAGKFKRPIIAGMEFSRSKSDNSSHPQKDEGDNPETIMSSGNVNSSS

Use ProtScale for the Chou-Fasman questions.

2. Check out ProteinSequence22B for different secondary structure elements.
    (Use a window size of 21.)

a. Check out the Chou-Fasman alpha-helical predictions. How many regions exceed a threshold of 1.1? (The threshold of 1.1 was chosen because results are only raw scores, and one should only consider strong peaks when interpreting graphs.)

b. Draw/show the Chou-Fasman alpha-helical plot below.

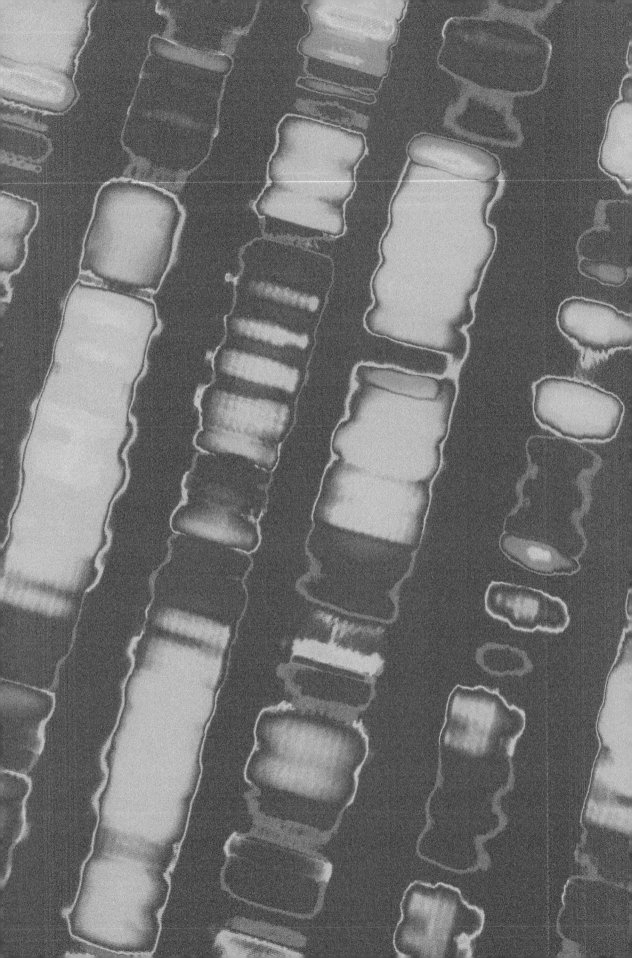

CHAPTER 03

SEQUENCE ALIGNMENT

lignments of biological sequences (DNA, RNA, and protein) have been a centerpiece of bioinformatics from its inception. Chapter 01 covered the rapid, high-performance pairwise BLAST aligner, and in this section we address a more sophisticated, albeit slower, method for generating pairwise sequence alignments that is guaranteed to produce a mathematically optimal alignment. We also discuss how pairwise alignments can be turned into multiple sequence alignments (MSAs).

What Is a Sequence Alignment?

A pairwise sequence alignment is a match of the ordered chemical letters between two different sequences. The goal of a sequence alignment is to vertically align the positions of the sequences that are homologous to one another (i.e., were derived from a common ancestor). For example, let's say I have two DNA sequences, one from a mouse and one from a human, both of which encode a very important protein sequence, namely globin proteins that are involved in binding or transporting oxygen (e.g., hemoglobin). Mice and humans are mammals. As such, they have inherited this gene from a common ancestor they shared long ago. Figure 3.1 shows the basic structures of the mouse and human globin proteins. The goal is to create a sequence alignment in which we properly align parts of the two different, but related, proteins with the same function.

Sequence Alignments: Nature's Experimental Results

DNA, RNA, and protein sequence alignments are fundamental to bioinformatics research. In addition to allowing rapid identification of molecular sequences (e.g., BLAST), sequence alignments underpin a majority of bioinformatics algorithms in one way or another. This is because MSAs of related molecular sequences from different organisms provide the equivalent of an experimental readout of millions or even billions of years of evolution. Instead of spending years manipulating every nucleotide or amino acid position in a gene to determine the functional consequences, one can examine an MSA full of related sequences to determine

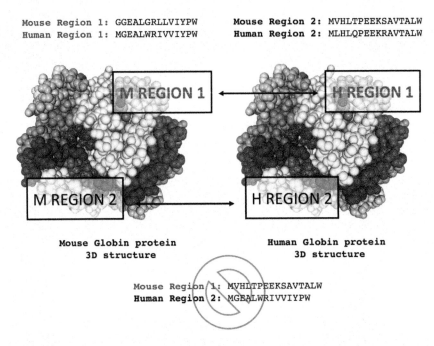

Mouse Region 1: GGEALGRLLVIYPW   **Mouse Region 2:** MVHLTPEEKSAVTALW
Human Region 1: MGEALWRIVVIYPW   **Human Region 2:** MLHLQPEEKRAVTALW

M REGION 1 ← → H REGION 1

M REGION 2 → H REGION 2

Mouse Globin protein   Human Globin protein
3D structure   3D structure

Mouse Region 1: MVHLTPEEKSAVTALW
Human Region 2: MGEALWRIVVIYPW

FIGURE 3.1. Structures of the mouse and human globin proteins. The bumpy blobs represent the three-dimensional structures of the mouse and human globin proteins. Notice how the structures are superimposable. The goal is to create a sequence alignment of the protein sequence, or underlying DNA sequence, in which the homologous functional regions of the proteins are aligned. In the process of making a sequence alignment between the mouse and human globin protein primary amino acid sequences, the sequence corresponding to mouse region 1 (M REGION 1) should be aligned to human region 1 (H REGION 1), and mouse region 2 should be aligned to human region 2. Clearly, it would make no sense to align the amino acids (or the underlying DNA that codes for these amino acids) of mouse region 1 to human region 2. This is an easy problem when the structures of the homologous molecules are available. The problem becomes harder when one has only the two sequences and no structure, and harder still the more distantly related (and different) the sequences are from one another.

which nucleotides (or amino acid positions) have experienced mutations and which have not.

To find the really important nucleotides or amino acids, any change of which would destroy the function of the macromolecule, here's a hint: look for the conserved alignment positions, the positions that have not changed in any of the sequences. The more constrained or conserved a nucleotide or amino acid sequence is, the more important it likely is for the functioning of the macromolecule. Mutations at these positions would almost certainly lead to the death of the organism, and natural selection would remove it from the gene pool. Figure 3.2 illustrates examples of sequence alignments showing both highly conserved positions and variable positions.

While the conserved positions in sequence alignments indicate functionally critical nucleotides or amino acids, the so-called variable positions can also be useful. The types and patterns of mutations allowed at these variable positions can be used in a large number of bioinformatics approaches that we will deal with in other chapters, including

- RNA structure prediction
- Motif searching
- Weight matrix construction
- Phylogenetic analysis
- Transition matrices

A

```
GTCAGGTCACAGTGACCTGAT
GTCAGGTCACAGTGACCTGAT
TGTAAGTCACTGTGACCTGTA
AGCAGGTCAAGGCGATCTGTA
   * ****    ** ***
```

B

```
FKGETQVTDQLTGY
VKGETQINDQLTGY
FKGETQVNDQLTGY
FKGETQVNDQLTGY
LRGETQISDQLTGY
.:****:.******
```

FIGURE 3.2. **MSAs of DNA and protein sequences indicating specific conserved and variable positions. (A)** MSA of a transcription factor binding site for the estrogen receptor protein. The underlined regions indicate the nucleotides that directly bind the protein transcription factor. Any mutation to the conserved nucleotides indicated by an asterisk would prevent the binding of the estrogen receptor transcription factor and prevent the transcription (and therefore translation) of a protein involved in fertility. **(B)** MSA of 5 related bacterial outer membrane protein sequences. The proteins are channels that allow ions to move in or out of bacterial cells. Asterisks indicate amino acids conserved in all the outer membrane proteins. The periods and colons indicate amino acid positions with less conservation.

- Protein secondary-structure prediction
- Hidden Markov models

What Are the Challenges in Aligning Sequences?

If sequences have identical regions, the sequence alignment problem is trivial: just slide the sequences alongside each other until you find a perfect match. However, when aligning sequences from different species, this is often not possible. As organisms become more distantly related, or if the rate of evolution of a particular gene is high, the sequences become more divergent and more difficult to align. While the sequences in the alignment perform more or less the same functional role in the different organisms, over time underlying DNA sequences diverge through the process of mutation. Nucleotides of the DNA sequence coding for the gene can be replaced (for instance, an A might change to a G), additional nucleotides can be inserted or deleted, and sometimes big stretches of the DNA can even be completely inverted. These changes occur for a variety of reasons. Errors in copying (genome replication[1]) of the sequence, errors in recombination, and mutagenic compounds can all cause mutations. If the mutations do not completely disable the gene and kill the organism, or prevent reproduction, these can be inherited; sometimes the mutations even prove advantageous.

This leaves the following problem, illustrated in Fig. 3.3: how to make an accurate sequence alignment between sequences with many differences that are not even of the same length.

The most difficult and important aspect of proper sequence alignments is putting gap characters in the right place. The gap characters represent insertions or deletions, also called indels (Fig. 3.3). The reason for using the term indel is that without knowing the history of the mutations, it is impossible to know if the gaps in the alignment represent an insertion in one sequence or a deletion in the other. To solve this problem, bioinformaticians invented types of dynamic programming algorithms that have scoring schemes for nucleotide (or amino acid) matches, mismatches, and gaps (indels). The activity section for this chapter explains how a dynamic programming approach uses scores for matches, mismatches, and gaps to determine the best alignment of two sequences. Generally, matches are preferred, so they get a positive score, while mismatches and gaps (also known as gap penalties) are to be avoided, so they get a zero or negative score. This maximizes the matches and minimizes mismatches or gaps, though it does not eliminate them.

Issues in Sequence Alignment

The biggest hurdle to accurate sequence alignment is the scoring system. With DNA sequences, the scoring system is fairly arbitrary: +1 for a match, −1 for a mismatch, and 0 for a gap is pretty common. However, these numbers can easily be changed. For instance, if the sequences are closely related, meaning that there has been less evolutionary time for the accumulation of insertion and deletion mutations which would create gaps in the sequence alignment, one could use a more negative gap penalty score. For closely related protein-coding DNA sequences, most default scoring schemes work very well. However, for distantly related sequences, it is often difficult to determine the proper gap penalties.[2]

A **Unaligned sequences**

        Human DNA: **ATGAATTTTAGTTTT**
        Fruit Fly DNA: **ATGATTAGTCTT**

B
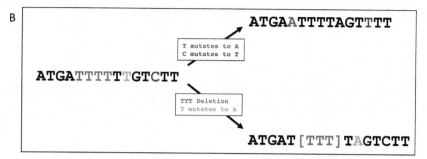

C **Aligned sequences**

        Human DNA: **ATGAATTTTAGTTTT**
        Fruit Fly DNA: **ATGATT---AGTCTT**

FIGURE 3.3. Mutational history of a protein-coding gene after evolution from a common ancestor. (A) The two unaligned sequences at the top are derived from a human and a fruit fly. These two sections of DNA code for a protein with the same cellular function, but they are clearly not identical. **(B)** The boxes in the middle show what mutations lead to the divergence in the sequences. The sequence on the left is the evolutionary ancestor of both the human and the fruit fly. The lineage leading to humans accumulated two substitution mutations (shown in red), while the lineage leading to fruit flies accumulated one substitution mutation (shown in green) and a deletion of three nucleotides (shown in blue brackets). **(C)** The final sequence alignment shows how these mutations lead to mismatches and how the deletion makes one sequence shorter than the other, which needs to be accounted for by gap characters when performing the alignment.

It turns out that the protein sequences themselves are much easier to align than the DNA sequences that code for the proteins. Not only is there less redundancy, but one can also use sophisticated scoring schemes for matches and mismatches. These scoring schemes are known as the PAM and BLOSUM matrices (see Chapter 07 for details on matrix calculation) and provide a score when an amino acid in one sequence matches the amino acid in the compared sequence and a mismatch score for every other amino acid. In protein sequence alignments mismatches can sometimes be useful, as we will discuss in Chapter 07. Activity 3.1 teaches how to use these scoring schemes to align protein sequences.

The most difficult types of sequences to align are noncoding RNA or DNA sequences. Unlike protein-coding genes, noncoding DNA (e.g., promoter regions upstream of the coding region) and DNA that encodes structural RNA can accumulate lots of insertions or deletions and still function. Insertions or deletions in protein-coding genes most often lead to frameshift mutations, which result in completely new and nonfunctioning protein sequences which are usually eliminated by natural selection. For RNA sequences, alignments are often performed by combining information on the RNA's structure with a dynamic programming method. Other methods, such as sliding window algorithms, can be used to align noncoding DNA.

Multiple-Sequence Alignment

In this chapter, we mainly focus on how to generate a sequence alignment for two sequences. In practice, one usually wants to create a MSA. It turns out that MSAs can be created by clever extension of pairwise sequence alignments. The progressive alignment method described by Paulien Hogeweg and Ben Hesper in 1984 was effectively integrated by the creators of the ClustalW program a decade later. The latest version of this software, one of the most-cited bioinformatics programs, is called Clustal Omega, which you will learn to use in Activity 3.1.

The basic approach of all the progressive alignment methods is as follows.

1. Make pairwise alignments of all the sequences.
2. Produce a guide tree based on the distances between all the pairs (see the distance method in Chapter 06 for an example).
3. Progressively align sequence pairs using the guide tree, starting with the most similar sequences.

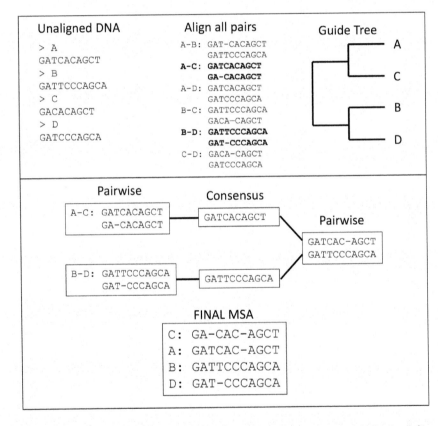

FIGURE 3.4. **Progressive sequence alignment of four DNA sequences.** At the top (left to right), beginning with the unaligned DNA sequences, all possible pairs are aligned using a dynamic programming approach. Distances are created from the sequence alignments and used to build a guide tree. At the bottom (left to right), using the guide tree, the most similar sequences are aligned in pairs. Consensus sequences are made from the pairs, and then the pair of consensus sequences is aligned. Note how the final MSA retains the gaps from the previous pairwise alignments as well as from the consensus pair alignment.

The key to this approach is that at each stage of the progressive alignment the program always aligns only two sequences. Figure 3.4 shows an example of a progressive MSA with four DNA sequences.

Notes

1. Retroviruses (such as HIV) have high rates of mutations that occur during genome replication and result in many mutant viral variants. Most result in nonfunctional viruses, but others result in new functioning variants of the virus that allow the virus to escape the immune system. Cool but evil, sort of like Darth Vader.

2. Good sequence alignments are vital for proper data interpretation. For years, poor sequence alignments that did not properly address large insertion or deletion mutations (gaps) led researchers to conclude that birds were related to mammals. This changed when a research group in China took a closer look and noticed that the sequence alignment was wrong.

ACTIVITY 3.1   DYNAMIC PROGRAMMING

Motivation

Alignments of different DNA, RNA, or protein sequences are a fundamental aspect of bioinformatics. Sequence alignments allow similarity comparisons (e.g., BLAST) for function prediction and organism identification. They also identify when mutations have occurred, such as the occurrence of single nucleotide polymorphisms in the human genome or the evolution of viruses during a pandemic. Multiple-sequence alignments are also critical for a number of other bioinformatics approaches, including many of the other bioinformatics tools covered in this book.

The purpose of this activity is to teach a dynamic programming algorithm for the global (end-to-end) alignment of any two DNA or protein sequences.[1] This algorithm, called the Needleman-Wunsch method, uses a scoring system and a grid matrix to efficiently determine the best alignment between a pair of DNA or protein sequences. Dynamic programming algorithms are common in mathematics, computer science, and bioinformatics. An algorithm is considered "dynamic programming" or "dynamic optimization" if it breaks down a problem into a set of easily solvable subproblems, each of which is stored and used for the eventual solution. (If only this would work with personal relationships.) The cool thing about these solutions is that, given a set of simple assumptions, they are guaranteed to produce the optimal sequence alignment.[2] In this case, the assumptions are a set of scores for the matching of two nucleotides, for the mismatching of two nucleotides, and for gaps when there has been an insertion or a deletion. This chapter will teach you how to solve dynamic programming problems for pairwise DNA and protein sequences and then teach you how to use an online program for constructing MSAs. The Needleman-Wunsch algorithm and dynamic programming can be a bit confusing at first glance, so make sure to read through the tutorial on the website carefully and practice using the interactives.

Learning Objectives

1. Learn the importance of sequence alignment and the challenges imposed by different types of mutations in generating an accurate sequence alignment (Motivation).
2. Use match, mismatch, and gap penalties to create a dynamic programming matrix for a pairwise DNA sequence alignment (Concepts and Exercises).
3. Use the matrix to perform the traceback step and determine the final, best global (end-to-end) alignment of two sequences (Concepts and Exercises).
4. Do the same for pairwise protein sequence alignment using the specialized protein scoring tables PAM and BLOSUM (Concepts and Exercises).
5. Learn how to use the Clustal Omega program to perform pairwise and multiple-sequence alignments (Concepts and Exercises).

Concepts

To prepare you to understand the principles and goals of dynamic programming, try the following anticipatory exercise. The Needleman-Wunsch method, and other pairwise sequence alignment methods, uses two-dimensional graphs to solve the alignment problem. Below you will find graphs that indicate two possible alignments of the same pair of DNA sequences, Seq1 and Seq2. Follow the paths to determine how to align the nucleotides (letters) of the two different sequences. Note that one sequence is longer than the other but the alignment is "global," which means that the sequences must be aligned from end to end. This also means that, if the sequences are of different lengths or very divergent from one another, you will need to adjust with spaces or other characters to create gaps in the sequence so that the correct bases line up properly. Most often hyphens are used to indicate these gaps (indels). Write the alignments below the graphs.

To begin the problem, in each graph start at the top left square and follow the X's. The X's indicate the path for each alignment. Notice how Seq1 is shorter than Seq2 but in the graph they are aligned end to end. This means that you must have gaps to get them to align properly. Both paths indicate that the G in Seq1 should be aligned to the G in Seq2. From there, both paths "move" diagonally and have the A in Seq1 aligned to the A in Seq2. However, the next X in the third square is diagonal in path 1 but vertical in path 2. This means that there are no bases in Seq1 to align with the G in Seq2, which creates a gap (indicated by a hyphen).

The first three positions of the alignments are given below. Try finishing the rest for both paths and answer the reflection questions.

Seq1: GATTTA

Seq2: GAGTTCA

PATH 1:

	G	A	T	T	T	A
G	X					
A		X				
G			X			
T			X			
T				X		
C					X	
A						X

PATH 2:

	G	A	T	T	T	A
G	X					
A		X				
G		X				
T			X			
T				X		
C					X	
A						X

Path 1:

Seq1: GAT _____

Seq2: GAG _____

Path 2:

Seq1: GA—_____

Seq2: GAG _____

Reflection

- How do you align the letters of the DNA when the path moves diagonally? Do the letters always match?
- What happens when the paths move vertically? Would a horizontal move be different and, if so, how?
- Could this type of graphing method be used for protein sequences?
- Which of the alignments seems better? Could some kind of scoring scheme help?

Below is the answer. Each path showed a different alignment for the same pair of sequences (Seq1 and Seq2). Because Seq1 is shorter, in an end-to-end alignment, all the letters have a match in Seq 1, but not all the Seq2 letters have a match. Thus, in order to get the sequences to align properly, it is necessary to put in some gaps (spaces) in the sequence alignment, indicated by a hyphen. The graph indicates what letters you should match up (a diagonal move is a match or a mismatch). However, when the path moves vertically, it stays on one sequence but moves along the other. This indicates that you must put a gap in the sequence alignment in the sequence that does not move. (At this point, I'm feeling quite moved.) Since in this case it is always Seq1, that is where all the gaps are placed. (A horizontal move would have been a gap in Seq2.)

Path 1:

```
Seq1: GAT-TTA
Seq2: GAGTTCA
```

Path 2:

```
Seq1: GA-TTTA
Seq2: GAGTTCA
```

So which of the two alignments for these same sequences is better? The second alignment path seems better, but we cannot know without a scoring scheme of some sort. Computers need numbers, and that is where the Needleman-Wunsch algorithm comes in.

Finding the path through the graph

Hopefully you know how to find the path through the graph and make a sequence alignment. (Though if you don't, keep reading.) But how do we determine the ideal path in the first place? This is where dynamic programming comes in. Figure 3.1.1 shows how to use dynamic programming for determining the best path through a two-dimensional graph. Note that in solving the problem, the algorithm calculates the values for just one cell at a time. That is the essence of dynamic programming: breaking a large problem into manageable subproblems. Then, after the scores for all the cells are determined, the algorithm traces back through the graph to find the best alignment. The Fig. 3.1.1 example uses the Needleman-Wunsch dynamic programming method to align a pair of short DNA sequences: ACT and ACA. A lame problem, admittedly, but Rome wasn't built in a day, was it? A critical aspect of the method is establishing a scoring scheme for matches, mismatches, and gaps. Given these scores, the method is guaranteed to provide the optimal solution. A typical scoring scheme might be something like the following.

<u>Scoring Scheme</u>: Match = +1; Mismatch = 0; Gap Penalty = –1

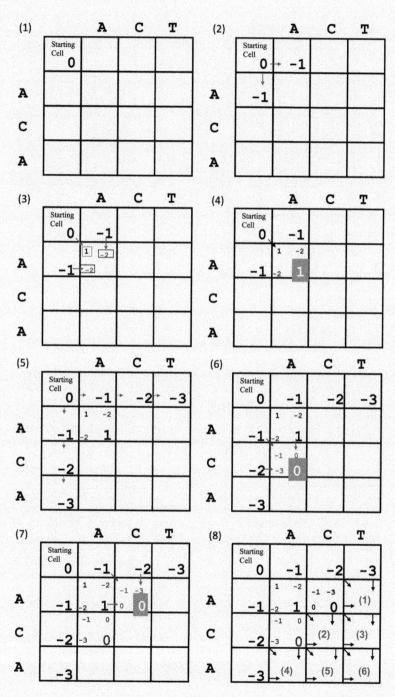

(continued)

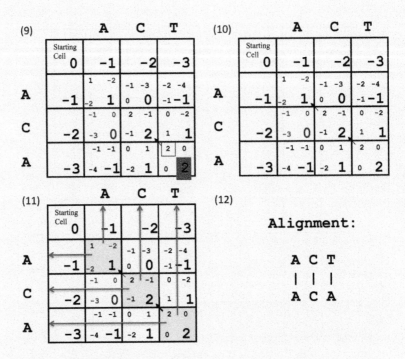

FIGURE 3.1.1. **Solving a dynamic programming alignment problem.** The value for each cell is based on the values from the surrounding cells. (1) Create a grid with your two sequences, adding an extra row and column (so, in this example, a 3-by-3 sequence becomes a 4-by-4 grid). Start with a 0 score in the top left corner. (Gotta start somewhere!) (2) Moving horizontally or vertically, use the gap penalty score. To fill in the top outside row, move horizontally and add the gap penalty to the score from the previous cell. For instance, moving right, add a −1 to the score from the original top left cell. Do the same to the leftmost column going down from the initial starting cell. (3) To determine the score for an empty cell, calculate the following three scores: the diagonal (match or mismatch), horizontal (gap penalty), and vertical (gap penalty), adding each score to that of the starting cell. In this case, the diagonal is a match (A to A), so we add +1 to the 0 that was in the top left cell, getting a diagonal score of +1. Next, determine the gap scores for the cell coming from the vertical and horizontal.[3] The gap penalty in this example is −1. The horizontal and vertical starting cells each have a score of −1, so the horizontal and vertical scores for the new cell are both −2. (4) Pick the highest (maximal) of the three numbers, and that is the final score for the cell, in this case +1. (5) The process repeats itself as you move to the next empty cell. (6) In this case, the diagonal is a mismatch (C to A), so we add 0 to the starting cell's score, giving us a −1 diagonal score. The vertical −1 gap penalty adds to the vertical starting score of 1, giving us a final vertical score of 0. The horizontal −1 gap penalty adds to the horizontal starting score of −2, giving us a final horizontal score of −3. So, the final score for the new cell is 0. (7) The process is repeated as you continue to fill out the rest of the empty cells until the entire graph is completed (8). (9) Solving the graph, traceback, and alignment. The Needleman-Wunsch algorithm creates a global (end-to-end) sequence alignment.[4] Once the graph is solved, start the traceback at the bottom right corner (the highest final score) for the global alignment. The arrows show the path of the traceback, which goes back to the cell from which the highest number was derived. In this case, the "2" is the highest, and it came from the diagonal cell. (10) Finish the traceback to the top left, and then (11) follow the path to make the alignment (12).

Exercises

Interactive exercise (theory)

Use the online Needleman-Wunsch alignment interactive link below to learn how to create scoring matrices, solve tracebacks, and produce pairwise sequence alignments. The Interactive Link explains how to use the teaching interactives. Once you learn how it works, solve the activity problem.

Needleman-Wunsch DNA/Protein Alignment Interactive Link

Link:

http://kelleybioinfo.org/algorithms /default.php?o=8

Problems

1. Fill in the blank cells using the Needleman-Wunsch algorithm and then complete the traceback. Using the traceback, write the sequence alignment in the spaces below.

Match : 2   **Sequence 1:** CTGTGG

Mismatch : -2   **Sequence 2:** CTGGG

Gap : 0   [Evaluate] [Reset] [Quiz] [Print Exam]

	S1	C	T	G	T	G	G
S2	0 0 / 0 0	0 0 / 0 0	0 0 / 0 0	0 0 / 0 0	0 0 / 0 0	0 0 / 0 0	0 0 / 0 0
C	0 0 / 0 0	2 0 / 0 2	-2 0 / 2 2	-2 0 / 2 2	-2 0 / 2 2		-2 0 / 2 2
T	0 0 / 0 0			0 2 / 4 4	4 2 / 4 4	0 2 / 4 4	0 2 / 4 4
G	0 0 / 0 0	-2 2 / 0 2		6 4 / 4 6	2 4 / 6 6	6 4 / 6 6	6 4 / 6 6
G	0 0 / 0 0	-2 2 / 0 2	0 4 / 2 4	6 6 / 4 6	4 6 / 6 6	8 6 / 6 8	8 6 / 8 8
G	0 0 / 0 0	-2 2 / 0 2	0 4 / 2 4	6 6 / 4 6	4 6 / 6 6		10 8 / 8 10

Sequence 1: [_____]

Sequence 2: [_____]

2. Fill in the blanks in this protein alignment using the Needleman-Wunsch algorithm and the PAM250 scoring matrix, shown in Fig. 3.1.2 (PAM250 is discussed further in Chapter 07). For this problem all the gap scores are −5, but the match and mismatch scores come from the PAM250 matrix. For example, an R-to-W mismatch according to the PAM250 matrix is +2, while an M-to-M match is +6. (Make sure that you can find these values in the PAM250 matrix.) To practice this type of problem, click on the Align Proteins button in the interactive module.

	A	R	N	D	C	Q	E	G	H	I	L	K	M	F	P	S	T	W	Y	V
A	2																			
R	-2	6																		
N	0	0	2																	
D	0	-1	2	4																
C	-2	-4	-4	-5	4															
Q	0	1	1	2	-5	4														
E	0	-1	1	3	-5	2	4													
G	1	-3	0	1	-3	-1	0	5												
H	-1	2	2	1	-3	3	1	-2	6											
I	-1	-2	-2	-2	-2	-2	-2	-3	-2	5										
L	-2	-3	-3	-4	-6	-2	-3	-4	-2	2	6									
K	-1	3	1	0	-5	1	0	-2	0	-2	-3	5								
M	-1	0	-2	-3	-5	-1	-2	-3	-2	2	4	0	6							
F	-4	-4	-4	-6	-4	-5	-5	-5	-2	1	2	-5	0	9						
P	1	0	-1	-1	-3	0	-1	-1	0	-2	-3	-1	-2	-5	6					
S	1	0	1	0	0	-1	0	1	-1	-1	-3	0	-2	-3	1	3				
T	1	-1	0	0	-2	-1	0	0	-1	0	-2	0	-1	-2	0	1	3			
W	-6	2	-4	-7	-8	-5	-7	-7	-3	-5	-2	-3	-4	0	-6	-2	-5	17		
Y	-3	-4	-2	-4	0	-4	-4	-5	0	-1	-1	-4	-2	7	-5	-3	-3	0	10	
V	0	-2	-2	-2	-2	-2	-2	-1	-2	4	2	-2	2	-1	-1	-1	0	-6	-2	4

FIGURE 3.1.2. **PAM250 scoring matrix.**

SCORE TABLE

PAM 250 : ◉
BLOSUM 62 : ○
Gap : -5

Sequence 1: MWVWM
Sequence 2: MRVWM

[Try Another] [Traceback]

S1	M	W	V	W	M
S2 0 0 0 **0**	0 0 0 **-5**	0 0 0 **-10**	0 0 0 **-15**	0 0 0 **-20**	0 0 0 **-25**
M 0 0 0 **-5**	6 -10 -10 **6**	-9 ☐ 1 ☐	-8 ☐ -4 ☐	-19 -25 -9 **-9**	☐ -30 -14 ☐
R 0 0 0 **-10**	-5 1 -15 **1**	☐ -4 -4 ☐	-1 -9 3 **3**	-2 -14 -2 **-2**	-9 ☐ -7 ☐
V 0 0 0 **-15**	-8 ☐ -20 ☐	-5 3 -9 **3**	12 -2 -2 **12**	-3 -7 ☐ ☐	0 -12 2 **2**
W 0 0 0 **-20**	-19 -9 -25 **-9**	13 -2 -14 **13**	-3 7 8 **8**	29 2 3 **29**	3 -3 24 **24**
M 0 0 0 **-25**	-14 -14 -30 **-14**	-13 ☐ -19 ☐	☐ 3 3 ☐	4 24 10 **24**	35 19 19 **35**

Lab Exercises (Practice)

In this part of the exercise, you will learn how to use the Clustal Omega program, which makes alignments of two or more sequences using dynamic programming algorithms.

Clustal Omega Tutorial Link
Link:
http://kelleybioinfo.org/algorithms/tutorial /TAli2.pdf

Sample and lab exercise data:
http://kelleybioinfo.org/algorithms /data/DAli2.txt

Lab Exercise

DNA multiple-sequence alignment

Unaligned DNA sequences for questions 1 to 3 below (and at the Sample and Lab Exercise data link):

>BR110MP90

TAATATCAATAGAAGAATTAGCCAAAATTACGTCCTGTCAAACCCCCTATGGTAAATAGAAAAATAAATC
CGATAGCTCATAGGGATGAAGGAGTTAAAGTAATTTGGGAGCCATGGTATGTTGCGAGTCATCTAAAAATTT
TGATTCCAGTAGGAACTGCAATAATTATTGTGGCTGATGTGAAGTAAGCTCGAGTATCAACATCTATCCCTACTGT
AAATATATGATGGGCTCACACTACAAATCCTAGCAGACCAATTGCTATTATAGCATAAATTATTCCTAATAAAC
CGAAAGCTTCCTTTTTGCCACTTTCTTGTCTAATAATATGAGAAATTATTCCGAAACCAGGTAAAATTAGAATAT
AAACTTCAGGATGTCCGAAAAATCAAAATAAATGCTGATAAAGAATAGGATCCCCTCCACCTGATGGG
TCAAAGAAGGTAGTATTAATATTTCGATCTGTCAATAGTATAGTAATAGCTCCGGCTAATAC

>JE05

AATAATATCAATGGAAGAATTGGCTAAAACTACACCAGTTAATCCCCCTAAAGTAAAGAGGAAAATAAAT
CCAATAGCTCAAAGGGAGGAGGGGTTTAGGGTAATTTGAGACCCATGGTATGTAGCTAACCATCTAAAAATTT
TGATTCCAGTCGGAACTGCAATAATTATTGTGGCAGACGTGAAATAGGCGCGAGTATCTACATCTATTCCTACTG
TGAACATATGATGGGCTCATACTACAAAACCTAATAGTCCAATTGCTATTATAGCATAAATTATTCCCAATAAT
CCAAAAGCTTCCTTTTTTCCTCTTTTCTTGCCTAATAATATGAGAAATTATACCAAATCCTGGTAAAATTAAAATAT
AAACTTCAGGGTGCCCAAAAAAATCAGAATAAGTGCTGATAGAGGATAGGGTCTCCTCCACCGGAGGGA
TCAAAAAAAGTAGTATTAATATTTCGGTCTGTTAAAAGTATAGTGATAGCCCCAGCTAAC

>JE06

GAATAATATCAATGGAAGAATTGGCTAAAACTACACCAGTTAATCCCCCTAAAGTAAAGAGGAAAATAAATC
CAATAGCTCAAAGGGAGGAGGGATTTAGGGTAATTTGAGACCCATGGTATGTAGCTAACCATCTAAAAATTT
TGATTCCAGTCGGAACTGCAATAATTATTGTGGCAGACGTGAAATAGGCGCGAGTATCTACATCTATTCCTACTGT
GAACATATGATGGGCCCATACTACAAAACCTAATAGTCCAATTGCTATTATAGCATAAATTATCCCCAATAATC
CAAAAGCTTCCTTTTTACCTCTTTCTTGCCTAATAATATGAGAAATTATACCAAATCCTGGTAAAATTAAAAATATA
AACTTCAGGGTGCCCAAAAAAATCAGAATAAGTGTTGATAGAGGATAGGGTCTCCTCCACCGGAGGGAT
CAAAAAAAGTAGTATTAATATTTCGGTCTGTTAAAAGTATAGTGATAGCTCCAGCTAA

>JE11L95

TATCAATGGAAGAATTGGCTAAAATTACACCAGTTAATCCCCCTAAAGTAAAGAGGAAAATAAATCCAATAGCT
CAAAGGGAGGAGGGAGTCAGGGTAATTTGAGATCCATGGTATGTAGCCAATCATCTAAAAATTTTAATTCCAGTC
CGAACTGCAATAATTATTGTGGCAGATGTAAAATAGGCGCGAGTATCTACATCTATTCCTACTGTGAAGTATATG
GATGAGCTCATACTACAAAACCTAATAATCCAATTGCTATTATAGCATAAATTATTCCTAATAATC
CAAAAGCTTCCTTTTTTCCTCTTTCTTGCCTAATAATATGGGAAATTATACCAAATCCTGGTAAAATTAAAATATA
AACTTCAGGGTGCCCAAAAAATCAAAATAAGTGTTGATAGAGGATAGGGTCTCCTCCACCGGAGGGGT
CAAAAAAAGTAGTATTAATATTTCGGTCTGTTAAAAGTATAGTGATAGCCCCAGCTAATACCG

>**JE17NB95**

AATATCAATGGAAGAATTGGCTAAAACTACACCAGTTAATCCCCCTAAAGTAAAGAGGAAAATAAATCCAATAGCT
CAAAGGGAGGAGGGATTTAGGGTAATTTGAGACCCATGGTATGTAGCTAACCATCTAAAAATTTTGATTCCAGTC
GGAACTGCAATAATTATTGTGGCAGACGTGAAATAGGCGCGAGTATCTACATCTATCCCTACTGTGAACATNAT
GATGGGCTCATACTACAAAACCTAATAGTCCAATTGCTATTATAGCATAAATTATTCCCAATAATC
CAAAAGCTTCCTTTTTTCCTCTTTCTTGCCTAATAATATGAGAAATTATACCAAATCCTGGTAAAATTAAAATATA
AACTTCAGGGTGCCCAAAAAATCAGAATAAGTGTTGATAGAGGATAGGGTCTCCTCCACCGGAGGGAT
CAAAAAAAGTAGTATTAATATTTCGGTCTGTTAAAAGTATAGTGATAGCCCCAGCTAACACC

>**JE19NB95**

ATATCAATGGAAGAATTGGCTAAAACTACACCAGTTAATCCCCCTAAAGTAAAGAGGAAAATAAATCCAATAGCT
CAAAGGGAGGAGGGATTTAGGGTAATTTGAGACCCATGGTATGTAGCTAATCATCTAAAAATTTTGATTCCAGTCG
GAACTGCAATAATTATTGTGGCAGATGTAAAATAGGCGCGAGTATCTACATCTATTCCTACTGTGAACATATGAT
GAGCTCATACTACAAAACCTAATAATCCAATTGCTATTATAGCATAAATTATTCCCAATAATCCAAAAGCTTCCTTT
TTAAACCTCTTTCTTGCCTAATAATATGAGAAATTATACCAAATCCTGGTAAAATTAAAATATAAACTTCAGGGT
GCCCAAAAAATCAGAATAAGTGTTGATAGAGGATAGGGTCTCCTCCACCGGAGGGATCAAAAAAAGTAGTATTA
ATATTTCGGTCTGTTAAAAGTATAGTGATAGCCCCAGCTAACACT

>**JE39M95**

TAATATCAATGGAAGAATTGGCTAAAACTACACCAGTTAATCCCCCTAAAGTAAAGAGGAAAATAAATC
CAATAGCTCAAAGGGAGGAGGGATTTAGGGTAATTTGAGACCCATGGTATGTAGCTAACCATCTAAAAATTTT
GATTCCAGTCGGAACTGCAATAATTATTGTGGCAGACGTGAAATAGGCGCGAGTATCTACATCTATCCCTACTGT
GAATATATGATGGGCTCATACTACAAAACCTAATAGTCCAATTGCTATTATAGCATAAATTATTCCCAATAATC
CAAAAGCTTCCTTTTTTCCTCTTTCTTGCCTAATAATATGAGAAATTATACCAAATCCTGGTAAAATTAAAATATA
AACTTCAGGGTGCCCAAAAAATCAGAATAAGTGTTGATAGAGGATAGGGTCTCCTCCACCGGAGGGAT
CAAAAAAAGTAGTATTAATATTTCGGTCTGTTAAAAGTATAGTGATAGCCCCAGCTA

1. Use the Clustal Omega alignment program with the default matrix to align the DNA sequence data. Write the first 10 positions of the alignment for only the first 4 aligned sequences below (or the first 4 rows of the alignment if you are using copy/paste).

2. This next section is designed to give you practice converting file formats. Many programs accept only a few selected file formats. Many allow you to use FASTA formats, but some need specialized formats, so practice converting file types is helpful. The conversion website used here functions similarly to the Clustal Omega site. Use **https://www.ebi.ac.uk/Tools/sfc/emboss_seqret/** to convert the Clustal alignment you created in question 1 into a FASTA format. (Note: use the ENTIRE file, including the header that begins with "CLUSTAL" and the asterisks at the bottom that are not actually part of the alignment.)

   Write the title lines and the first 10 positions of the alignment for only the first 3 sequences in the FASTA file (or the first 3 sequences in the alignment if you are using copy/paste).

3. Convert the Clustal alignment to the Nexus/paup interleaved format, which is used often in phylogenetic analyses.

   a. What kind of information does the header of the Nexus/paup file contain?

   b. What do you think "ntax" and "nchar" refer to?

   c. What is at the very last line of the file?

Protein multiple-sequence alignment

Unaligned protein sequences for question 4 below (and at the Sample and lab exercise data link):

>LCseedSfl

MKKLTVAISAVAASVLMAMSAQAAEIYNKDSNKLDLYGKVNAKHYFSSNDADDGDTTYVRLGFKGETQINDQLTG
FGQWEYEFKGNRAESQGSSKDKTRLAFAGLKFGDYGSIDYGRNYGVAYDIGAWTDVLPEFGGDTWTQTDVFM
TGRTTGVATYRNNDFFGLVDGLNFAAQYQGKNDRTDVTEANGDGFGFSTTYEYEGFGVGATYAKSDRTNDQVIY
GNNSLNASGQNAEVWAAGLKYDANNIYLATTYSETQNMTVFGNNHIANKAQNFEVVAQYQFDFGLRPSVAYLQSK
GKDLGAWGDQDLIEYIDVGATYYFNKNMSTFVDYKINLIDKSDFTKASGVATDDIVAVGLVYQF

>PhoEseedEco1

MKMKKSTLALVVMGIVASASVQAAEIYNKDGNKLDVYGKVKAMHYMSDNDSKDGDQSYIRFGFKGETQINDQL
TGYGRWEAEFAGNKAESDTAQQKTRLAFAGLKYKDLGSFDYGRNLGALYDVEAWTDMFPEFGGDSSAQTDNFM
TKRASGLATYRNTDFFGVIDGLNLTLQYQGKNENRDVKKQNGDGFGTSLTYDFGGSDFAISGAYTNSDRTNEQNLQ
SRGTGKRAEAWATGLKYDANNIYLATFYSETRKMTPITGGFANKTQNFEAVAQYQFDFGLRPSLGYVLSKGKDIEGI
GDEDLVNYIDVGATYYFNKNMSAFVDYKINQLDSDNKLNINNDDIVAVGMTYQF

>PhoEseedEco2

MKMKKSTLALVVMGIVASVSVQAAEIYNKDGNKLDVYGKVKAMHYMSDNDSKDGDQSYIRFGFKGETQINDQL
TGYGRWEAEFAGNKAESDTAQQKTRLAFAGLKYKDLGSFDYGRNLGALYDVEAWTDMFPEFGGDSSAQTDNFM
TKRASGLATYRNTDFFGVIDGLNLTLQYQGKNENRDVKKQNGDGFGTSLTYDFGGSDFAISGAYTNSDRTNEQNLQ
SRGTGKRAEAWATGLKYDANNIYLATFYSETRKMTPISGGFANKTQNFEAVAQYQFDFGLRPSLGYVLSKGKDIEGI
GDEDLVNYVDVGATYYFNKNMSAFVDYKINQLDSDNKLNINNDDIVAVGMTYQF

>PhoEseedEco4

MKKSTLALVVMGIVASASVQAAEIYNKDGNKLDVYGKVKAMHYMSDNDSKDGDQSYIRFGFKGETQINDQLT
GYGRWEAEFAGNKAESDTAQQKTRLAFAGLKYKDLGSFDYGRNLGALYDVEAWTDMFPEFGGDSSAQTDNFM
TKRASGLATYRNTDFFGVIDGLNLTLQYQGKNENRDVKKQNGDGFGTSLTYDFGGSDFAISGAYTNSDRTNEQNLQ
SRGTGKRAEAWATGLKYDANNIYLATFYSETRKMTPITGGFANKTQNFEAVAQYQFDFGLRPSLGYVLSKGKDIEGI
GDEDLVNYIDVGATYYFNKNMSAFVDYKINQLDSDNKLNINNDDIVAVGMTYQF

>PhoEseedSen1

MNKSTLAIVVSIIASASVHAAEVYNKNGNKLDVYGKVKAMHYMSDYDSKDGDQSYVRFGFKGETQINDQLT
GYGRWEAEFAGNKAESDSSQQKNRLAFAGLKLKDIGSFDYGRNLGALYDVEAWTDMFPEFGGDSSAQTDNF
MTKRASGLATYRNTDFFGIVDGLDLTLQYQGKNEDRDVKKQNGNGFGTSVSYDFGGSDFAVSGAYTLSDRTREQNLQ
RRGTGDKAEAWATGVKYDANDIYIATFYSETRNMTPVSGGFANKTQNFEAVIQYQFDFGLRPSLGYVLSKGKD
IEGVGSEDLVNYIDVGATYYFNKNMSAFVDYKINQLDSDNTLGINDDDIVAIGLTYQF

>PhoEseedSen2

MNKSTLAIVVSIIASASVHAAEVYNKNGNKLDVYGKVKAMHYMSDYDSKDGDQSYVRFGFKGETQINDQLT
GYGRWEAEFAGNKAESDSSQQKTRLAFAGLKLKDIGSFDYGRNLGALYDVEAWTDMFPEFGGDSSAQTDNFM
TKRASGLATYRNTDFFGIVDGLDLTLQYQGKNEDRDVKKQNGDGFGTSVSYDFGGSDFAVSGAYTLSDRTREQNLQ
RRGTGDKAEAWATGVKYDANDIYIATFYSETRNMTPVSGGFANKTQNFEAVIQYQFDFGLRPSLGYVLSKGKD
IEGVGSEDLVNYIDVGAIYYFNKNMSAFVDYKINQLDSDNTLGINDDDIVAIGLTYQF

>PhoEseedSfl

MKKSTLALVVMGIVASASVQAAEIYNKDGNKLDVYGKVKAMHYMSDNASKDGDQSYIRFGFKGETQINDQLT
GYGRWEAEFAGNKAESDTAQQKTRLAFAGLKYKDLGSFDYGRNLGALYDVEAWTDMFPEFGGDSSAQTDNFM
TKRASGLATYRNTDFFGVIDGLNLTLQYQGKNENRDVKKQNGDGFGTSLTYDFGGSDFAISGAYTNSDRTNEQNLQ
SRGTGKRAEAWATGLKYDANNIYLATFYSETRKMTPITGGFANKTQNFEAVAQYQFDFGLRPSLGYVLSKGKDIEGI
GDEDLVNYIDVGATYYFNKNMSAFVDYKINQLDSDNKLNINNDDTVAVGMTYQF

>PhoEseedSty

MNKSTLAIVVSIIASASVHAAEVYNKNGNKLDVYGKVKAMHYMSDYDSKDGDQSYVRFGFKGETQINDQLT
GYGRWEAEFASNKAESDSSQQKTRLAFAGLKLKDIGSFDYGRNLGALYDVEAWTDMFPEFGGDSSAQTDNFM
TKRASGLATYRNTDFFGIVDGLDLTLQYQGKNEDRDVKKQNGDGFGTSVSYDFGGSDFAVSGAYTLSDRTREQNLQ
RRGTGDKAEAWATGVKYDANDIYIATFYSETRNMTPVSGGFANKTQNFEAVIQYQFDFGLRPSLGYVLSKGKD
IEGVGSEDLVNYIDVGATYYFNKNMSAFVDYKINQLDSDNTLGINDDDIVAIGLTYQF

>nmpCseedEco1

MNIYRAVTSFFNNSSKKGLTMKKLTVAISAVAASVLMAMSAQAAEIYNKDSNKLDLYGKVNAKHYFSSNDADD
GDTTYARLGFKGETQINDQLTGFGQWEYEFKGNRAESQGSSKDKTRLAFAGLKFGDYGSIDYGRNYGVAYDIGAWT
DVLPEFGGDTWTQTDVFMTQRATGVATYRNNDFFGLVDGLNFAAQYQGKNDRSDFDNYTEGNGDGFGFSATYEYE
GFGIGATYAKSDRTDTQVNAGKVLPEVFASGKNAEVWAAGLKYDANNIYLATTYSETQNMTVFADHFVANKAQN
FEAVAQYQFDFGLRPSVAYLQSKGKDLGVWGDQDLVKYVDVGATYYFNKNMSTFVDYKINLLDKNDFTKEGANKSLI

>nmpCseedSty

MKLKLVAVAVTSLLAAGVVNAAEVYNKDGNKLDLYGKVHAQHYFSDDNGSDGDKTYARLGFKGETQINDQLTG
FGQWEYEFKGNRTESQGADKDKTRLAFAGLKFADYGSFDYGRNYGVAYDIGAWTDVLPEFGGDTWTQTDVFMT
GRTTGVATYRNTDFFGLVEGLNFAAQYQGKNDRDGAYESNGDGFGLSATYEYEGFGVGAAYAKSDRTNNQVKAA
SNLNAAGKNAEVWAAGLKYDANNIYLATTYSETLNMTTFGEDAAGDAFIANKTQNFEAVAQYQFDFGLRPSIAYLKS
KGKNLGTYGDQDLVEYIDVGATYYFNKNMSTFVDYKINLLDDSDFTKAAKVSTDNIVAVGLNYQF

4. Use Clustal Omega with the unaligned protein sequences.

   a. Find the longest stretch of <u>complete conserved positions</u>. Write the amino acids for this stretch of sequence below (which is the same in all the sequences).

   b. Find the region or regions of the alignment in which most or all of the sequences have a 3-character gap ("---"). Write the amino acid letters for the sequences that do NOT have gaps.

Notes

1. This is also called a pairwise sequence alignment.
2. BLAST is faster and highly accurate, but it is not guaranteed to produce an optimal sequence alignment.
3. It is necessary to have the scores from the three surrounding cells, which is why we added an extra row and column when we started the graph.
4. The Smith-Waterman variant of this algorithm produces a local alignment, which aligns two sequences but does not constrain the alignment to be from end to end. This is important when aligning, say, a short sequence to an entire genome. In the Smith-Waterman variant, all the negative cell values are changed to 0 when the graph is calculated, which makes the positive local alignments visible. Then the traceback starts at the highest scoring cell regardless of where that value is in the graph, and proceeds until it reaches a 0 value.

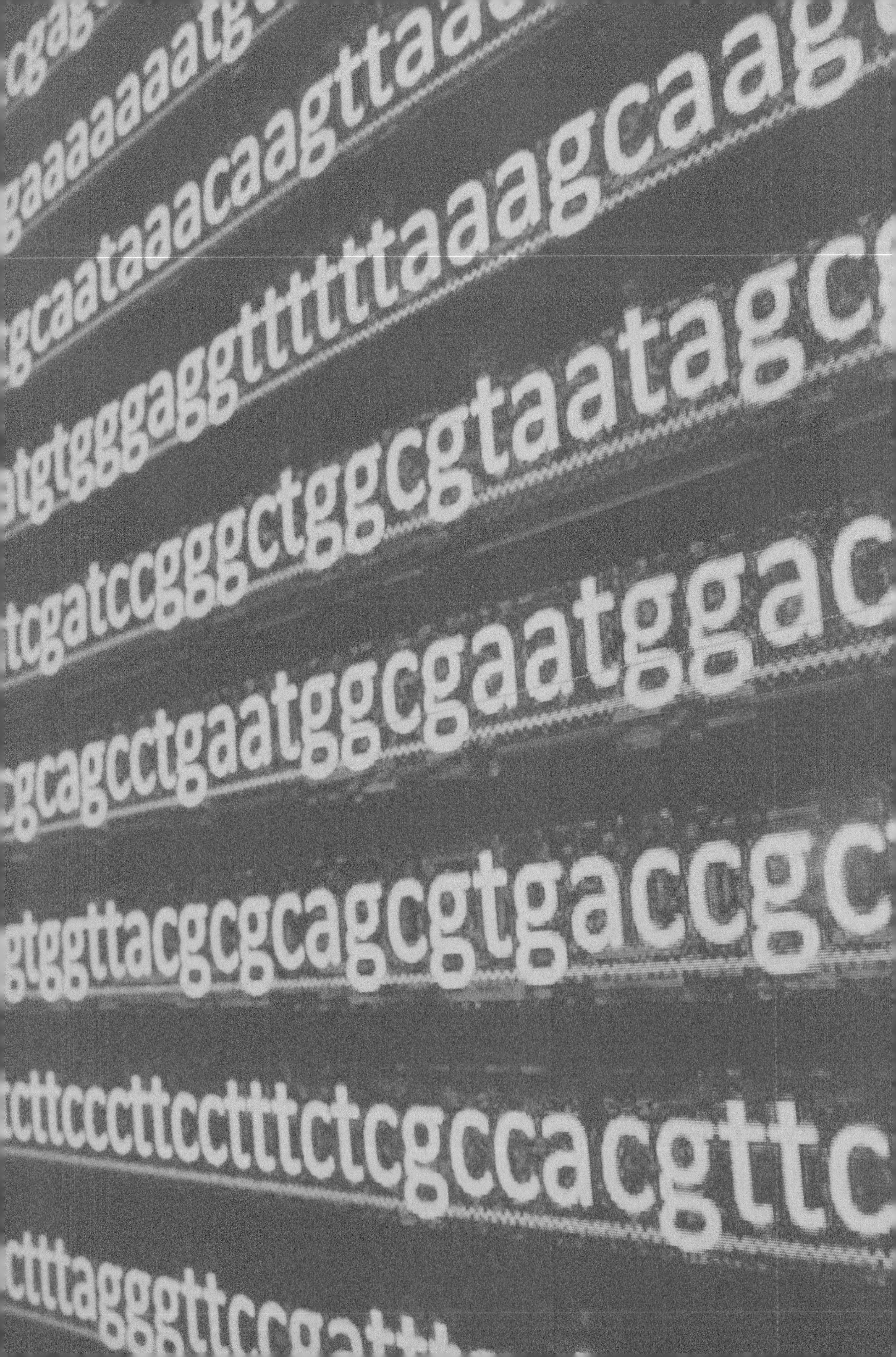

gaaaaacaagtta aagcaag

gcaataaacaagtta aagcaag

gtgggagttttt taatagc

tcgatccgggctggcgtaatagc

gcagcctgaatggcgaatggac

gtggttacgcgcagcgtgaccgc

tcttcccttcctttctcgccacgttc

ctttaggggg

PATTERNS IN THE DATA

arly in the primordial days of molecular sequencing, analysis of DNA and protein sequence alignments uncovered interesting nonrandom patterns of nucleotide variation. One of the earliest discoveries came from the comparison of the upstream (5′) promoter regions of protein-coding genes, right in the midst of the binding site of the RNA polymerase. This region contained what is known as a TATA box, which is vital for the correct binding of the RNA polymerase and transcription of the messenger RNA. Analysis of the sequence alignment containing the TATA box region, even visually, shows a significant bias towards T and A nucleotides (Fig. 4.1). The TATA box turns out to be crucial for binding the protein known, shockingly enough, as the TATA binding protein, which is a critical part of the RNA polymerase complex. Experimental mutations of these conserved nucleotides reduced or completely eliminated the process of transcription. If such a mutation happened in a gene that was critical for organism development or survival, the organism with the mutation would be eliminated via natural selection and the mutation would not pass on to future generations. Harsh, but true.

Such patterns proved to be extremely common across a vast variety of genomes. Figure 4.2 illustrates a number of other examples using sequence logos. Sequence logos were created by Tom Schneider and Mike Stephens back in the dark ages (1990, when cell phones were the size of bricks and email was only for nerds) and were designed as a simple visual means of illustrating the highly conserved positions in a sequence alignment (DNA, RNA, or protein). These positions were likely to be extremely important and, therefore, conserved by natural selection. The calculation of sequence logos is quite simple and is based on information theory. The level of conservation at a position of a sequence alignment, R_{seq}, is calculated as follows:

$$R_{seq} = \log_2 N - \left(-\sum_{n=1}^{N} p_n \log_2 p_n \right)$$

where N is the number of distinct symbols, and p_n is the observed frequency of symbol n at position p. For DNA or RNA, there are four symbols—A, G, C, and T

```
tacgaccccaTTTAGTagtcaaccgc
gtttcgccccTATTACagactcacaa
cattgtgaatTAATATgcaaataaag
attgaataccTATTATgagttagcca
acgctttacgTATAGTggcgacaatt
atcgcaactcTCTACTgtttctccat
atccagtaacTATAAAagcatatcgc
tgtagaatttTATTCTgaatgtgtgg
gtgggctctcTATTTTaggattaatt
tcctcctgttTATTCTtattaccccg
```

FIGURE 4.1. Sequence alignment of regions 5′ upstream from 10 protein-coding genes in the bacterial *Escherichia coli* genome. There is a clear bias towards T and A nucleotides, though none of the sequences are completely identical.

(U for RNA)—and the maximum sequence logo score for a position is 2 if all the nucleotides at a position are the same[1] (e.g., all A nucleotides). Protein sequences have many more symbols, 20 total representing each of the most frequently occurring amino acids, so the maximum bit score (if 100% of the amino acids at a position are of one type) is 4.32.

Figure 4.2 illustrates sequence logos made for sequence alignments of non-coding DNA sequences.

Similar conserved patterns can be detected in protein sequence alignments. These regions indicate amino acids that are critical in the functioning of the protein. For instance, highly conserved amino acids in a transcription factor (TF) could be critical in allowing the TF DNA-binding domain to bind DNA.

Sequence Motifs

The types of short sequence patterns shown in the figures here can also be found in RNA and protein sequences, and are generally referred to as motifs. One goal of bioinformatics has been to develop algorithms to automatically identify functional motifs in the vast sea of DNA, RNA, and protein databanks. For instance, we might want to search for all the alternate splice sites in the genome of the fruit fly or find all the binding sites of a particular TF in the genome of the Black Death bacterium.

The algorithms covered in this chapter are designed to take into account the facts that motifs tend to be short and that, while certain positions tend to be conserved, there is also a lot of positional variability among related motifs. The two methods we discuss in the chapter include (i) a basic protein sequence motif algorithm and (ii) a DNA motif search tool called a weight matrix.[2] In general, these and other motif-searching algorithms have the following in common.

1. They are based on experimental data.
2. They find motifs too short for BLAST searches.
3. They find motifs too "fuzzy" for BLAST searches.
4. They allow "weighted" positional bias.
5. They generate functional hypotheses.

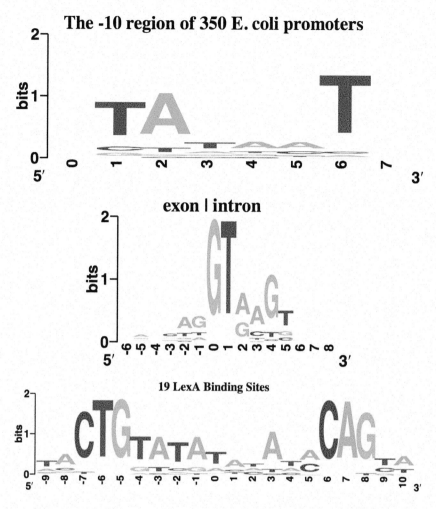

FIGURE 4.2. Logos for sequences involved in critical cell functions. The bigger the letter is shown at a position, the more conserved (and important) that nucleotide is in the process. **(Top)** Conserved sequences within *E. coli* promoter regions. **(Middle)** Exon and intron splice sites. **(Bottom)** A site in a bacterial genome that binds a TF. The sequence logos were generated using the online WebLogo software (**http://weblogo.berkeley.edu/**).

That these motifs are based on experimental data cannot be overstated. They must be built on alignments of experimentally tested sequences of known function, and the more the better. Since the motifs themselves tend to be short and have highly variable positions as well as conserved positions, one generally cannot use BLAST to find matching motifs in new sequences. However, because they are short and fuzzy, matches using motif-searching algorithms need to be taken as hypothetical because many are likely to be false positives.

Notes

1. This is also known as the number of "bits."
2. Chapter 07 discusses the general method called hidden Markov models, which can also be used to find motifs.

ACTIVITY 4.1   PROTEIN SEQUENCE MOTIFS

Motivation

Protein sequence motifs are short stretches of amino acids with specific functional roles that are found in many different types of proteins. For instance, the DNA-binding motif known as a zinc finger is found in numerous different types of transcription factors (TFs), proteins that regulate the transcription of DNA to messenger RNA. Although the amino acids in the binding motif are similar across many different TFs, the other amino acids in these proteins can be very different, yet they all need to bind DNA; hence, they have a DNA-binding motif. Being able to identify motifs, given the primary protein sequence, can provide important insight into a functional aspect of unknown proteins.

In this activity, you will learn how to turn a multiple-sequence alignment of known protein motifs into a search pattern for scanning new proteins for the same motif. In the first step, you will learn how to build a position-specific search pattern using the sequence alignment. In the second step, you will learn how to scan new protein sequences for positive matches. You will also learn how to use online motif searching software to search for matching motifs in protein sequences.

Learning Objectives

1. Understand the basics of sequence motifs and how detecting motifs in proteins helps to identify important functional aspects (Motivation).
2. Learn how to build a sequence motif from alignments of short protein sequences and use them to search for matches within novel protein sequences (Concepts and Exercises).
3. Use sequence motif matching software to build and search for protein motifs (Concepts and Exercises).

Concepts

This preliminary exercise will help you understand the principles behind protein sequence motif searching. The alignment on the next page contains a series of short sequences of a protein motif in the DNA-binding domain of the androgen receptor (AR) from various different mammal species. The AR binds the hormone testosterone, and once activated by the hormone binding, the AR moves from the cell cytoplasm to the nucleus, where it binds the regulatory DNA sequences of many genes. This binding helps to activate the transcription of genes which are vital for male reproduction and development.

			Positions					
			1	2	3	4	5	6
Human	**AR**	**Motif:**	R	V	L	E	G	Q
Dog	**AR**	**Motif:**	R	A	M	E	G	K
Camel	**AR**	**Motif:**	R	A	M	E	G	Q
Horse	**AR**	**Motif:**	R	V	M	E	G	K
Mouse	**AR**	**Motif:**	R	V	V	E	G	Q
Bear	**AR**	**Motif:**	R	A	L	E	G	K

Look at each position of the alignment. Can you make a pattern (known as a profile) that matches all the sequences? Hint: the first letter in the pattern would be an R because they all have an arginine (R) at the first position.

PROFILE:

FIND PROFILE MATCH IN THIS SEQUENCE:

M F W V Y R V M E G K S K

Reflection
- Which positions are the most conserved? Most variable?
- How might you search for your pattern in a database of protein sequences? (How did the BLAST algorithm find the first matching "word"?)
- What if there were 10 possible choices of amino acid at position 2? How might you indicate a hypervariable position?

Below is the profile for this sequence alignment and a match to the profile in the test protein sequence. The answer uses hyphens to indicate separate positions of the motif.

PROFILE: R - [V or A] - [L or M or V] - E - G - [Q or K]

PROFILE MATCH IN THIS SEQUENCE:

M F W V Y <u>R V M E G K</u> S K

In three of the positions of the sequence alignment, all the motifs have an identical amino acid. This suggests that mutating any of these three amino acids would inhibit the binding of the AR to its target DNA sequence. For instance, if one were to mutate the DNA sequence that coded for this motif so that there was a D at position 4 instead of an E, the motif would probably no longer bind the correct DNA sequence.

One especially bad consequence of such a mutation would be infertility. Since the AR is vital in the process of spermatogenesis (development of sperm), individuals with such a mutation could not reproduce and this mutation would not pass to the next generation.[1] This process of natural selection is the likely reason why one only observes an E at the 4th position of this motif in all these species (or an R at the 1st, or a G at the 5th). Once a profile is generated for a particular sequence motif, this profile can then be scanned across millions of proteins in databases for matches.

Exercises

Interactive exercises (theory)
Use the online exercise link below to learn how to make protein sequence motifs and use them to search for matches. The Interactive Link explains how to use the teaching interactive. Once you learn how it works, solve the activity problem.

Sequence Motif Interactive Link
Link:
http://kelleybioinfo.org/algorithms /default.php?o=4

Problem

Build the sequence motif and indicate whether the profile you built matches the protein sequences below.

Protein 1	S	C	T	S	C	M
Protein 2	S	Q	T	S	C	M
Protein 3	S	M	N	S	C	Q
Protein 4	S	M	F	S	C	M
Protein 5	S	C	F	S	C	Q
Protein 6	S	M	N	S	C	Q
Protein 7	S	C	N	S	W	M

Motif : ☐ - ☐ - ☐ - ☐ - ☐ - ☐

Which protein sequence matches the profile? **YES NO**

Sequence 1	S	Q	N	F	C	Q	○ ○
Sequence 2	S	M	N	S	L	M	○ ○
Sequence 3	S	M	T	S	C	M	○ ○
Sequence 4	S	Y	N	S	C	Q	○ ○
Sequence 5	S	Q	T	S	C	M	○ ○

Lab Exercises (Practice)

In this part of the exercise, you will learn how to use sequence motifs to search for matches online. You will also learn how to interpret the output from the program.

ScanProsite Tutorial Link
Link:
**http://kelleybioinfo.org/algorithms
/tutorial/TMot2.pdf**

Sample and lab exercise data:
**http://kelleybioinfo.org/algorithms
/data/DMot2.txt**

Lab Exercise

1. Make a profile manually from the following data using the pattern syntax used by PROSITE. See tutorial and link for ScanProsite.

   Protein sequences for generating a profile:

P2_DROME/200-226	CVVCGDKSSGKHY
DR_CANFA/547-573	CLICADEASGCHY
1H16_CAEEL/14-40	CAICQESAEGFHF
2H18_CAEEL/11-37	CEVCPDKTSYRHF
3H18_CAEEL/11-37	CPVCGDRTSLRHF
4H18_CAEEL/11-37	CPPCGDLTSPSHF
5H18_CAEEL/11-37	CDLCGDPSRGWHF
F6H18_CAEEL/11-37	CFQCLDWTAGANF
7H18_CAEEL/11-37	CTWCPDQTGWFHF
8H18_CAEEL/11-37	CWPCWDPTVGGHY
9H18_CAEEL/11-37	CEACGDKTLGYHF
0H18_CAEEL/11-37	CSFCGDKTIPRNF
AH18_CAEEL/11-37	CPRCQEDTQRYHY
BH18_CAEEL/11-37	CQECGDKTWWRNF

   a. What is the profile?

   b. Use the Option 2 "Submit MOTIFS" to search for hits with your profile in the PROSITE database (use the default parameters). Fill in the table below with the Swiss-Prot/UniProtKB accession numbers of hits to three different organisms, as well as the motif sequence from the organism that matches the profile you created, and the scientific and, if available, common name of the organism. (The "Pattern" should be the sequence from the organism that matches your search pattern.)

Swiss-Prot/UniProtKB Acc #	Pattern (Specific Match)	Organism
i.		
ii.		
iii.		

2. Use ScanProsite Option 1 to scan the protein P04150 (UniProtKB Identifier) and answer the following questions.

   a. What is the protein?

   b. What motif(s) did you find (i.e., what amino acid sequence or sequences)?

   c. Describe the motif and what it binds.

3. Search the P04150 using InterProScan at **http://www.ebi.ac.uk/interpro /search/sequence-search**. Select the Advanced Options menu below the input window and select only 2 boxes: PfamA and Prosite-Profiles. What major domains did you discover?

4. Use the SMART (Simple Modular Architecture Research Tool) to study your cool new sequence from a feral Chernobyl chicken[2] (use the normal, not the genomic, version).
SMART home page: **http://smart.embl-heidelberg.de/**.

> **Chernobyl chicken**

MQFAPLLLGVFLLCGSARGSDSSASNAITCFTRGLDLRKETEDVLCPANCPLWQFYVFGDGI
YASLSSVCGAAIHRGVITNAGGAVRVQTLPGQENYPAVHANGIQSQVLSRWASSFSVTPGTN
NLALEAVGRSVATARPATGKRPKKTLEKKAGNKDCKADIAFLIDGSYNIGQRRFNLQKNFV
GKVAVMLGIGTEGPHVGVVQASEHPKIEFYLKNFTAAKEVLFAIKELGFRGGNSNTGKALK
HAAQKFFSMENGARKGIPKIIVVFLDGWPSDDLEEAGIVAREFGVNVFIVSVAKPTTEELGM
VQDIGFIDKAVRCRNNGFFSYQMPSWFGTTKYVKPLVQKLCSHEQMLCSKTCYNSVNIGFLI
DGSSSVGESNFRLMLEFISNVAKAFEISDIGSKIATVQFTYDQRTEFSFTDYTTKEKVLSAIRNI
RYMSGGTATGDAISFTTRNVFGPVKDGANKNFLVILTDGQSYDDVRGPAVAAQKAGITVFS
VGVAWAPLDDLKDMASEPRESHTFFTREFTGLEQMVPDVINNNGICKDFLDSKQ

   a. Search for outlier homologues, PFAM, signal peptides, and internal repeats. What high-probability motifs did you discover? Can you describe them in words?

   b. What is the name/function of the protein? (Hint: BLAST the protein sequence.)

5. Use Ensembl at **http://www.ensembl.org** to answer the next few questions about the homolog of the Chernobyl chicken protein in the mouse genome. (Hint: use the Ensembl Tutorial in the BASICS section and search the mouse genome.)

    a. Which mouse chromosome is it on?

    b. How many exons does it have?

    c. What is the length of the longest mRNA transcript in nucleotides? Ensembl uses bp (base pairs) because it is based on the DNA, but RNA does not have pairs since it is single stranded, so it should be nucleotides (nt).

Notes

1. Natural selection requires both survival AND reproduction, not to mention variation and heritability. Sequence alignments are a powerful way to determine the results of nature's experimentation: what is/is not allowed in nature.
2. These feral chickens probably fed on the gamma radiation-eating fungi inside the reactor core: https://www.sciencedaily.com/releases/2007/05/070522210932.htm

ACTIVITY 4.2   POSITION-SPECIFIC WEIGHT MATRICES

Motivation

Transcription factors (TFs) help to determine whether a particular gene (or set of genes) is transcribed into messenger RNA. So-called activator TFs help activate (increase) the transcription of a gene. Most activator TFs bind DNA upstream (5′) of the protein-coding region and make it more likely that RNA polymerase will bind and transcribe the DNA template strand. RNA polymerase might bind anyway, but activators make it more likely to happen by direct or indirect interactions with the RNA polymerase. Repressor TFs also bind DNA but do the opposite, making it less likely that RNA polymerase will transcribe the DNA. TFs typically bind to specific sequences of DNA, although how stringent these sequences are can vary. In fact, they usually bind to different, though highly similar, DNA sequences. In order to predict these DNA binding sites, we need a method that takes this variation into account. This was the idea behind the creation of the position-specific weight matrix (PSWM).

This activity teaches how to create PSWMs using sequence alignments of experimentally determined TF binding sites (TFBSs) and how to use a PSWM to search and find new sequences that match the PSWM. These matches could be binding sites for the TF. This is similar in principle to the creation of protein sequence motifs, but with DNA sequences. It's the same idea, but more math-y. After calculating the PSWM, you will learn how to scan DNA sequences for high-scoring matches to the PSWM and then use online software for this purpose.

Learning Objectives

1. Understand the principles of PSWMs and how they are used for detecting protein-DNA binding sites (Motivation).
2. Be able to construct and calculate a PSWM and use it to scan a DNA sequence for significant matches (Concepts and Exercises).
3. Learn how to use PSWM software to detect DNA binding sites of TFs (Concepts and Exercises).

Concepts

This preliminary exercise should help you understand the principles behind PSWMs. The sequence alignment on the next page contains a series of short DNA sequences that have been experimentally determined to bind the TF known as HotStuff (Donna Summer's BFF TF).

Binding site sequences for the HotStuff TF

```
1   2   3   4   5   6
— — — — — — — — — —
A   G   C   T   A   A
T   G   C   T   G   A
A   G   C   T   C   G
T   G   C   T   T   G
A   G   T   T   A   A
T   G   C   T   G   G
A   G   C   T   T   A
T   G   C   T   C   A
— — — — — — — — — —
```

How many nucleotides are at each position? (Fill in the table below.)

<u>Position</u>

	1	2	3	4	5	6
A	4					
G						
C						
T						

What is the best match in the following sequence?
(Hint: the alignment has 6 positions.)

ACAATGCTCAAGGG

Reflection
- Which positions are the most conserved? The most variable?
- Is position 3 more weighted towards a C or a T? How would you describe the weight of position 1?
- How might you score a match to this PSWM using the frequency of the nucleotides at each position?
- How many of each nucleotide should we see at each position if all bases were equally likely? Hint: there are 4 nucleotides in DNA, and assume each is equally common (1/4 A, 1/4 G, 1/4 C, and 1/4 T).

On the next page are the answers showing the number of each nucleotide at every position and the best match of the matrix in the sequence. Notice how the positions are weighted towards particular nucleotides at certain positions. Position 1 is weighted equally between A and T but away from G and C, while position 3 is heavily weighted towards C, though there is a slight possibility of a T. There is no weight towards any of the nucleotides at position 5—all are equally likely.

	1	2	3	4	5	6
A	4	0	0	0	2	5
G	0	8	0	0	2	3
C	0	0	7	0	2	0
T	4	0	1	8	2	0

To get the best match in a sequence, we scan the sequence from left to right, looking at six nucleotides at a time. Why six? Because the alignment is six nucleotides long. The winner is the one with the best overall match to the matrix.

ACAATGCTCAAGGG

The question then arises: how do we get a score for the best fit to the matrix? In order to answer this question, we need one more transformation of the matrix. In this transformation, the frequency of the nucleotides is used to determine a score for each nucleotide at each position. The matrix transformation is essentially the natural log of the frequency of the base at each position divided by the expected frequency of that base at that position.

For example, the observed frequency of A at position 1 is 0.5 (50%, or 4 out of 8 possible nucleotides). At position 3, C has an observed frequency of 0.875 (87.5%, or 7 out of 8 possible nucleotides). Clearly, these are more common than expected by chance. If there were no biases, we would expect a frequency closer to 0.25 (25%, or 2 out of 8) of each nucleotide at each position. Calculating the PSWM score is easy. Simply calculate the natural log of the likelihood of the base at that position,[1] which is the observed frequency (f) of each nucleotide (i) at each position (j) divided by the expected frequency of each nucleotide (p_j):

$$\ln \frac{f_{i,j}}{p_i}$$

We will assume that the bases are all present at equal frequencies in the organisms. Since DNA has four nucleotides, each base has a p_i of 0.25. $p_a = p_g = p_t = p_c = 0.25$. (This is not always the case, since some organisms are G/C rich while others are A/T rich, but close enough!) This equation works until you realize that many of the frequencies in the table are 0, and you cannot take the natural log of 0. However, with a few adjustments to the equation, we can calculate a very close approximation:

$$\ln \frac{\left(n_{i,j} + p_i\right)/(N+1)}{p_i}$$

where $n_{i,j}$ is the number of bases i at position j and N is the total number of sequences in the alignment. The numerator is very close to the frequency, but adding the extra 0.25 ensures that the number is never 0.

Using our equation, we can then transform the following matrix of position-specific nucleotide counts to a PSWM. Let's start with position 1.

	1	2	3	4	5	6
A	4	0	0	0	2	5
G	0	8	0	0	2	3
C	0	0	7	0	2	0
T	4	0	1	8	2	0

Calculating the PSWM score for the A at position 1 is as follows. Since $n_{A,1} = 4$ (counts of A's at position 1), $p_i = 0.25$, and $N = 8$ (number of sequences), we get the following:

$$\ln \frac{(4 + 0.25)/(8 + 1)}{0.25} = +0.64$$

This is the same value for T at position 1, and the 0s in the matrix all have the same value:

$$\ln \frac{(0 + 0.25)/(8 + 1)}{0.25} = -2.19$$

Using the PSWM equation (and a calculator, unless you're a math savant), we can readily fill out the PSWM values for position 1 and the rest of the positions (the highest PSWM score(s) at each position are indicated in bold):

	1	2	3	4	5	6
A	**+0.64**	−2.19	−2.19	−2.19	**0.0**	+0.85
G	−2.19	**+1.29**	−2.19	−2.19	**0.0**	**+0.37**
C	−2.19	−2.19	**+1.17**	−2.19	**0.0**	−2.19
T	**+0.64**	−2.19	−0.59	**+1.29**	**0.0**	−2.19

Exercises

Interactive exercises (theory)

Now that you know the basics of how to calculate a PSWM using a sequence alignment, the online exercises will give you practice calculating matrices and teach you how to scan sequences to find a region with the highest score given the matrix. The Interactive Link has a more detailed explanation of how to calculate a PSWM and how to use it to scan a sequence looking for the best match. Once you learn how it works, solve the activity problem.

Weight Matrix Interactive Link
Link:
http://kelleybioinfo.org/algorithms/default.php?o=11

Problem

Enter the correct values in the boxes below (except the ones in the equation). First, fill in the number of each of the four nucleotides at position 3. Second, use these numbers to calculate the PSWM values for position 3. Finally, fill in the position-specific scores for the nucleotides in the highlighted 5-base sequence window and calculate the total score.

WEIGHT MATRIX

C	T	G	T	A
C	G	G	A	T
A	T	A	T	T
C	T	A	T	G
A	T	A	T	T
T	T	A	T	T
C	T	T	T	T
C	T	A	T	T
A	T	A	T	T
C	T	A	T	T

Number of Columns (5 - 10)

`5`

Sequence Length (max = 50 bases)

`10`

SUBMIT   RESET   PRINT EXAM

Number of nucleotides at each alignment position

A	3	0		1	1
C	6	0		0	0
G	0	1		0	1
T	1	9		9	8

$$Weight = Ln \frac{(n_{i,j} + p_i)/(N+1)}{p_i}$$

$$= Ln \frac{(\boxed{0} + \boxed{0})/(\boxed{0} + 1)}{\boxed{0}}$$

$= \boxed{0}$ COMPUTE

Weight Matrix

A	0.17	-2.40		-0.79	-0.79
C	0.82	-2.40		-2.40	-2.40
G	-2.40	-0.79		-2.40	-0.79
T	-0.79	1.21		1.21	1.10

T	C	G	C	A	T	A	C	G	C

Score:

Lab Exercises (Practice)

In this part of the exercise, you will learn how to use a program that searches sequences for TFBSs. This program uses PSWMs, like the ones you created earlier, to detect potential binding sites in new sequences.

Transcription Factor Binding Site Tutorial Link
Link:
**http://kelleybioinfo.org/algorithms/tutorial
/TMot1.pdf**

Sample and lab exercise data:
**http://kelleybioinfo.org/algorithms/data/
DMot1.txt**

Lab Exercise

>EP_1

GAGAGCGGGCAGGAGGCGGGTTGGGAGGGCGCGGAGCCCCGGGTTCGGGGGAGACTGGAG
GGGCGCACGTGCGGCCGGGTGCGAGCGCGCGGCGGGGGAGGCTGCGGGGCGGCGCGGGGG
CGCGCGCGGAGCCCGAGCGGCGGCGCCAGGTCACACAACCTGTTTTGGCGCCTGCGGGCG
CCTGGGCCCAAGGGTGCGACGCGGGGGGCGCCTGAGCCGGGACACAGGGGGTGCGGTGAGC
GCCAGGCGCCGCGGGGAGTTAAAAAGTTCGGGACCTGAGCGGTGCGTGGTTCCGCGGTGG
CCGCCTCTTCCTGCCGCGCAGGCCGAGGGTCCCGACGGCGCCGCTCACCGCTCCGGGACT
CAGCCTTTCTGGGCCCGGCCTGCGGTTCCCTCGGGGCCGGGGAGAGGGTGGAGCGCGGGA
GGAGGGGCGCCGGGTGGGGACGCCCAGGCCCTTCGTCGGGGGAGGGCGCTCCACCCGGGC
TGGAGTTGCAGAGCCCAGCAGATCCCTGCGGCGTTCGCGAGGGTGGGACGGGAAGCGGGC
TGGGAAGTCGGGCCGAGGTGGGTGTGGGGTTCGGGGTGTATTTCGTCCACGAGCCGGGGA

Use the sequence above with the LASAGNA TFBS search program to answer the following questions. (Under "Matrix-Derived Models", choose "Use TRANSFAC Matrices.")

1. What are the names and scores of the TFs with the two highest-scoring matrices (the two largest numbers in the Score column)?

2. What is the sequence snippet within EP_1 sequence (Sequence column) that matches the second-highest-scoring TF and therefore could be its binding site?

3. In the results, select the name of the TF in the left column. Using the References link, can you describe a known biological role of the highest-scoring TF?

Notes

1. Why calculate the log likelihood? Using the natural log (or logarithms) of a number makes calculations more convenient, especially with large numbers. For instance, instead of multiplying frequencies, you can add the logs of these numbers.

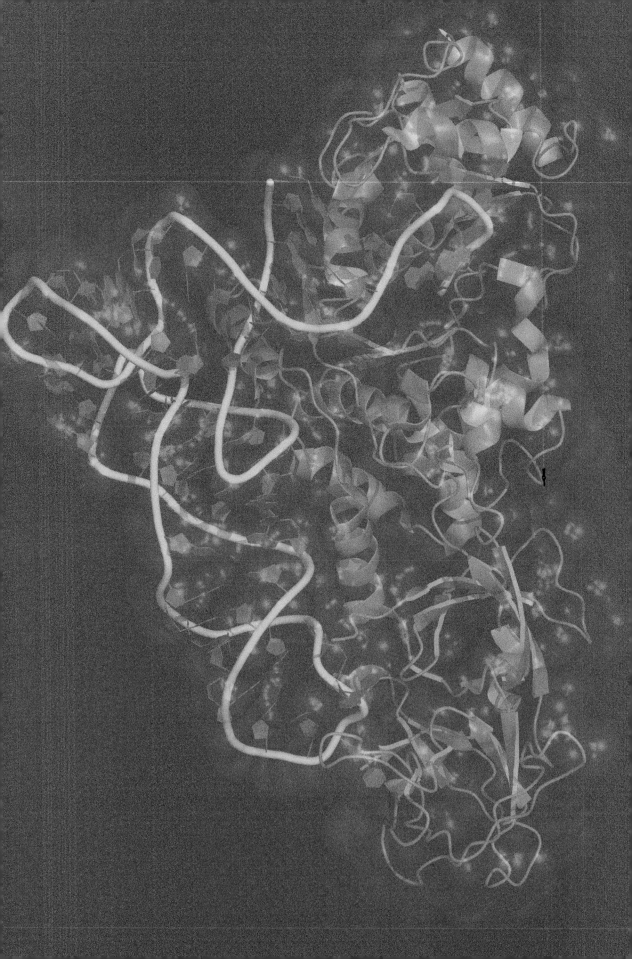

RNA STRUCTURE PREDICTION

NA is fundamentally boring, biochemically speaking. Sure, DNA contains the blueprint for the entire structure and function of all living cells (big deal), but the double helix is as stable and inactive a molecule as they get. The two antiparallel strands combine via their complementary Watson-Crick base pairings, with multiple hydrogen bonds per pairing. Without lowering the energy of activation using protein helicase enzymes, one must bring DNA to around 95°C (nearly the boiling point of water) in order to separate the strands of the double helix. Indeed, this stability is part of what makes DNA such a robust transmitter of biological information. DNA can be unwrapped, unzipped, copied, and transcribed, and it then re-forms quickly into a perfect double helix.

DNA's cousin RNA, on the other hand, is an entirely different story. The chemical composition of RNA is very similar to that of DNA, but with a few changes that make all the difference. Like DNA, RNA is a long polymer chain composed of four repeating nucleotides in which the sugar components of the nucleotides are joined together by phosphodiester bonds. RNA has a pentose sugar slightly different from that of DNA (ribose versus deoxyribose), with an additional hydroxyl group. RNA also features the nitrogenous base uracil in place of thymine. Most importantly, as shown in Fig. 5.1, cellular RNA molecules do not form double helices.

The single-stranded nature of RNA makes it a particularly dynamic and interesting molecule and gives RNA the potential for both structural flexibility and biochemical activity. Because only one strand of RNA is synthesized, the nucleotides are not bound to their complements on another strand, leaving the hydrogen bonds open. This allows them to form interactions with proteins and other RNA molecules and, most importantly, with themselves. Given the right sequence of nucleotides, and sometimes a little help from proteins, RNA molecules can form a variety of complex structures that perform critical cellular functions, as depicted in Fig. 5.2.

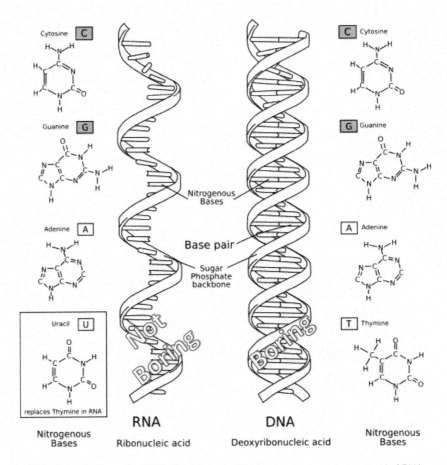

Cytosine **C**

Guanine **G**

Adenine **A**

Uracil **U**

replaces Thymine in RNA

Nitrogenous
Bases

RNA

Ribonucleic acid

Nitrogenous
Bases

Base pair

Sugar
Phosphate
backbone

DNA

Deoxyribonucleic acid

C Cytosine

G Guanine

A Adenine

T Thymine

Nitrogenous
Bases

FIGURE 5.1. RNA versus DNA. The left side of the image shows a single strand of RNA, as well as the chemical structures of the four nitrogenous bases that it comprises. The right side shows a double-stranded DNA helix and its four nitrogenous bases. Credit: NHGRI/Darryl Leja.

Roles of RNA in Cells

You already have some familiarity with structural RNA molecules from learning about the process of transcription and translation (see Chapter 00). For example, during translation, transfer RNA (tRNA) molecules (Fig. 5.2A) shuttle amino acids to the ribosome, the protein "factory" of cells. The ribosome itself is also mostly composed of RNA.[1] The two subunits of the ribosome, called the small and large subunits, each have structural RNA molecules called the small subunit RNA and large subunit RNA, respectively, that form the core of this protein-RNA molecular complex (Fig. 5.2B).

tRNAs, because they are relatively small, critical to cellular function, and highly abundant, were the first crystallized RNA structures. The three-dimensional rendering of this molecule clearly revealed how RNA structures form: by the RNA folding upon itself. Figure 5.2 shows examples of RNA structure folds, in which RNA molecules use hydrogen bonds to bind to themselves, making antiparallel

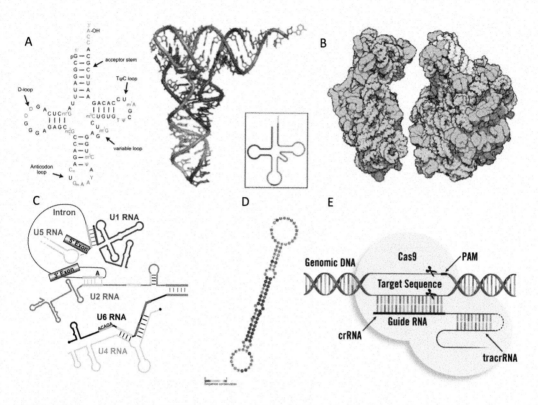

FIGURE 5.2. Some sweet, sweet RNA molecules and a few selected molecular interactions. (A) A tRNA molecule. tRNAs are made from a single strand of RNA which base-pairs with itself to form a structure capable of carrying amino acids to the ribosome. **(B)** Ribosomal subunits are composed of RNA (yellow and orange) and proteins (blue) that come together to form the ribosome, the cell's protein-synthesizing machinery. **(C)** The spliceosome, made of small nuclear RNAs and proteins, is part of the RNA editing machinery in eukaryotic cells. **(D)** Secondary structure of the mir210 microRNA. MicroRNAs regulate the expression of other genes. **(E)** The CRISPR-Cas9 system is a genome editing system found in prokaryotic cells to protect against bacteriophage invaders, which uses RNA at multiple steps to guide the process. Panels A and B courtesy of Yikrazuul, under license CC BY-3.0. Panel C reprinted from Will CL, Luhrmann R. 2011. *Cold Spring Harb Perspect Biol* 3(7), with permission. Panel E courtesy of Mirus Bio LLC. http://www.mirusbio.com

structures. Figure 5.2A shows a two-dimensional flattened secondary structure of a tRNA that indicates how the nucleotides complement each other, much like in DNA. The paired regions are called stems, and the RNA must loop around to fold back upon itself. These loop regions can have very interesting biochemical properties. The unpaired nucleotides in these loops have open hydrogen bonds that can interact with and bind other molecules, or even bind other regions within the same molecule. For instance, the anticodon hairpin loop of tRNAs (Fig. 5.2A) has open hydrogen bonds that allow the anticodon nucleotides of a particular tRNA to fleetingly bind mRNA during translation. This binding process is critical for attaching the correct amino acid during protein synthesis.

Figure 5.3 shows an example of a larger and more complex structural RNA, ribonuclease P (RNase P), the first naturally occurring RNA molecule proven to

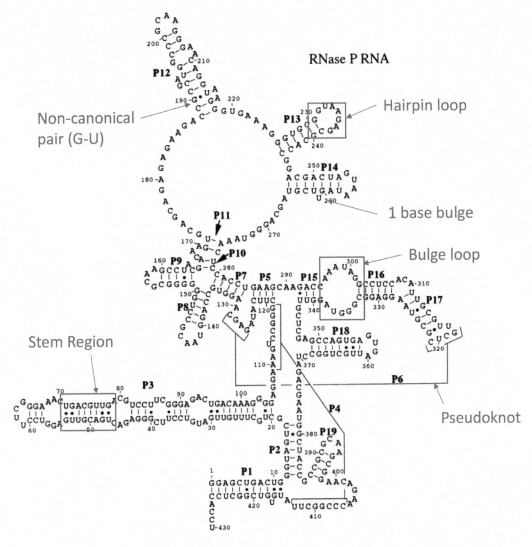

FIGURE 5.3. Secondary-structure diagram of the RNase P structural RNA showing examples of RNA structural elements known to occur in this and other molecules. Adapted from Maeda T, Furushita M, Hamamura K, Shiba T. 2001. FEMS Microbiol Lett 198:141–146, with permission.

have enzymatic activity in the absence of a protein component. The secondary-structure diagram illustrates a variety of structural elements found in the molecule. Many of these elements were predicted by bioinformatics methods and later verified experimentally in the laboratory. These types of structural elements are also found in many other RNA molecules and together determine both the secondary and much of the tertiary structure of the molecule.

Predicting RNA Structure

So much DNA, so little time! That should really be the subtitle of this book. RNA molecules, like protein sequences, are also encoded in all organisms' DNA, and just like with proteins, there are a mess of them. While many of the large structural RNAs have been well characterized, there are thousands of smaller RNAs predicted to be in genomes that remain to be studied. With the rapid and accelerating accumulation of DNA sequences, the question is: can we use DNA sequences that code for potential structural RNA molecules to predict their secondary or even tertiary structures?

The two methods we will cover for predicting RNA structure based on primary sequence information are (i) thermodynamics-based prediction and (ii) mutual information (MI). Thermodynamic methods use experimentally determined RNA base-pairing and base-stacking values to determine what the most stable potential RNA structure is for an RNA sequence. Given a DNA sequence encoding a structural RNA, these methods first determine a series of potential ways in which the sequence could fold upon itself. Then, using the experimentally predetermined RNA stacking energies, the algorithm determines the free energy of all the structures, and the one with the lowest free energy wins (like golf, but more exciting). Figure 5.4 illustrates potential folds and energies for the same RNA sequence.

Since folding algorithms maximize the number of base-pairings, and assume that the molecule folds only upon itself with lots of stems, thermodynamics-based predictions tend to perform more poorly with larger RNA structures. Also, the larger the structure, the more possible folds the RNA could make, which

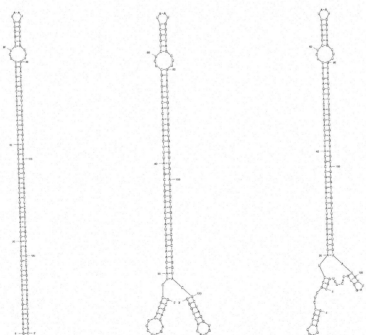

FIGURE 5.4. Three RNA folds for the same sequence. Fold 1 has the lowest free energy and would be preferred by the thermodynamic prediction method. Reprinted from Maeda T, Furushita M, Hamamura K, Shiba T. 2001. *FEMS Microbiol Lett* 198:141–146, with permission.

exponentially increases the computational time. These methods also do not predict higher-level interactions such as pseudoknots and base triples. However, thermodynamic methods can be extremely fast, require only a single sequence to make predictions, and tend to work well with small RNA molecules.

The second method, MI, addresses the problem of RNA secondary-structure prediction very differently. Instead of folding a single RNA sequence by itself, MI uses evolutionary information from many closely related sequences. Specifically, MI analyzes the mutational variation in multiple-sequence alignments generated from collections of the same RNA structure from many different organisms. In these alignments, MI identifies variable positions in the sequence alignment in which changes (mutations) at one position of the sequence alignment correlate with changes at another position. These correlated positions mutually inform one another, hence the name. The concept is that if two positions (or more in some cases) always change at the same time, this provides evidence that the positions are interacting within a sequence. Figure 5.5 provides an example of how and why correlated mutations like this occur.

The example in Fig. 5.5 is actually quite a common pattern. Paired regions often show strong patterns of correlated mutations because stem regions and pseudoknot regions are likely to be critically important to the stability and function of the molecule. Too many uncompensated mutations will lead to a poorly functioning molecule and could lead to death of the organism. In other words,

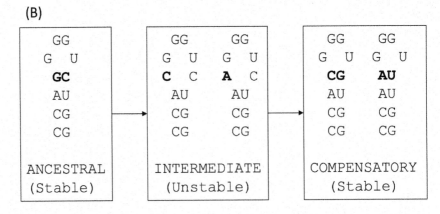

FIGURE 5.5. **Example of compensatory mutations in RNA structure. (A)** Multiple-sequence alignment of three related RNA sequences with the same function but from different organisms. The two positions indicated in boldface are correlated. **(B)** The structures below the alignment show the process of compensatory mutation. In this case, the second (compensatory) mutation restores the stability of the RNA molecule disrupted by the first (intermediate) mutation because base pairs in stem regions have a lower free energy than in hairpin loops.

natural selection would take its course. For instance, mutations that destabilize the stems of a tRNA molecule could lead to a molecule that does not function in protein synthesis. No protein synthesis means no proteins, which means no cell.

Since MI cares only about correlated mutations in a sequence alignment, more than just base pairs in stem regions can be detected. This means that pseudoknots and other strange interactions, such as base triples, can be determined given enough sequence data and variation. Clearly, the downside of MI is the need for lots of sequences and also a significant amount of variation among the sequences (no variation means no correlation or prediction). However, with the growing sequence databases from thousands of new genomes, sequence data are not really a limiting factor anymore.

Notes

1. The ribosome is a complex RNA-protein macromolecule.

ACTIVITY 5.1   RNA STRUCTURE PREDICTION

Motivation

Most of the RNA diversity in cells comes in the form of messenger RNA (mRNA) destined for the ribosome, where the mRNA is used in protein synthesis and later recycled. However, there are also many critical structural RNA molecules that have a variety of basic cellular functions and are not transcribed into proteins. In fact, the ribosome itself is largely made of two structural RNAs, one in the small subunit and one in the large subunit. Structural RNA molecules called transfer RNAs (tRNAs) shepherd the amino acids to the ribosome during protein synthesis. Other structural RNAs are involved in gene regulation and intron splicing, and some even act as enzymes.

As with proteins, predicting the structure of these RNAs helps us understand how they function. This activity covers two algorithms for predicting RNA structure. The first uses thermodynamic folding rules to find the best two-dimensional (secondary structural) fold of a given RNA sequence. RNA strands easily fold on themselves and form hydrogen bonds, making helical-like regions similar to DNA. The second method uses the principles behind mutual information (MI). MI requires alignment of different RNA sequences and looks for instances when mutations are correlated to predict likely-interacting RNA nucleotides. After mastering the principles of these algorithms, you will learn how to use online RNA thermodynamic prediction and MI prediction software and how to interpret their output.

Learning Objectives

1. Learn about the biological function of structural RNA molecules (Motivation).
2. Understand both principles of RNA folding and secondary and tertiary structural elements (Motivation).
3. Use free-energy thermodynamic rules to choose the most stable RNA fold among a series of folds for the same RNA sequence (Concepts and Exercises).
4. Learn the principle of MI and how it can be used to predict both secondary and tertiary RNA structural elements (Concepts and Exercises).
5. Learn how to use RNA prediction software and interpret the output (Concepts and Exercises).

Concepts

Algorithm 1: thermodynamic secondary-structure prediction

Problem: Given a set of possible RNA secondary structural folds for a particular RNA sequence, which is the best?

Solution: To answer this question, we will use the thermodynamic principle of free energy. The RNA structure with the lowest total free energy, i.e., the most stable predicted structure, will be chosen as the best prediction.

To better understand the principles behind the thermodynamic structure prediction method, try the preparatory exercise below. Using your brain and a pencil, try folding the RNA sequence upon itself by bringing complementary nucleotides (A and U, G and C) together. Start by drawing lines connecting paired nucleotides, and then draw the structure similar to the ones shown in Fig. 5.5. (Hint: start by connecting the second nucleotide and the last nucleotide.)

Sequence: **5′ – G A G G U C G G A A G A C C U – 3′**

Structure:

Reflection

- What types of RNA elements does your fold have?
- How many Watson-Crick pairings did you find? How many unpaired nucleotides?
- If you were to mutate the fifth nucleotide from a U to a G, how would this change the structure? What would this do to the stability of the molecule?
- Since A-U or U-A pairs have 2 hydrogen bonds, and G-C or C-G pairings have three, how many total hydrogen bonds are there in your structure?

Below is the folded structure. Since the nucleotides are all connected from 5′ to 3′ via phosphodiester (covalent) bonds, the molecule must twist around to be able to fold on itself. This is what creates the hairpin loop structure.

Sequence: **5′ – G A G G U C G G A A G A C C U – 3′**

Structure:

```
    G  A              G  A
    G     A           G     A
    C  G              C  G
    U  A              G     A
    G  C              G  C
    G  C              G  C
    A  U              A  U
    G                 G

  BEST FOLD        SWITCH U to G at position 5
```

The best fold has a stem and a hairpin loop structure. The total number of hydrogen bonds for this fold is 13. It is the combination of these hydrogen bonds plus the stacking energy that determines the free energy of the stem regions. The other main determinants of the molecule's free energy are the number and size of the loop regions, which raise the free energy of the molecule and lower its stability. For instance, the mutation from a U to a G in the 5th position of the sequence results in the creation of an internal 2-base loop because the nucleotides no longer pair. It is easy to see how this addition would significantly raise the free energy of the RNA structure.

Calculating RNA free energy

Thermodynamic methods use empirically (experimentally) determined free energies of RNA structural elements, including nucleotide pairings (A-U, U-A, G-C, C-G, and even G-U[1] and U-G), stacking energies, hairpin loops, internal loops, and bulges. Specifically, these methods use what are known as the nearest-neighbor rules, shown in Fig. 5.1.1.

		TOP					
		AU	**CG**	**GC**	**UA**	**GU**	**UG**
B O T T O M	**AU**	-0.9	-1.8	-2.3	-1.1	-0.5	-0.7
	CG	-2.1	-2.9	-3.4	-2.3	-1.5	-1.5
	GC	-1.7	-2	-2.9	-1.8	-1.3	-1.5
	UA	-0.9	-1.7	-2.1	-0.9	-0.7	-0.5
	GU	-0.9	-1.7	-2.1	-0.9	-0.5	-0.5
	UG	-0.9	-1.7	-2.1	-0.9	0.6	-0.5

Bases in Loop	Internal Loop	Bulge Loop	Hairpin Loop
1	0	3.3	0
2	0.8	5.2	0
3	1.3	6	7.4
4	1.7	6.7	5.9
5	2.1	7.4	4.4
6	2.5	8.2	4.3
7	2.6	9.1	4.1
8	2.8	10	4.1

FIGURE 5.1.1. **Nearest-neighbor free-energy rules for RNA structures, first developed by Turner and colleagues in 1999 and then updated in 2004.**[2] The first (uppermost) table shows the free energies for base pairs stacked over other base pairs. For instance, an A-U base pair stacked over a C-G base pair (A-U and C-G are "neighbors" in which the A and C and the U and G are covalently bound) has a total free energy of −2.1. This free energy includes both the energy of the pairing (A to U) and the fact that it is next to a C-G pair. The lower table shows the free energies of hairpin loops, internal loops, and bulge loops. Notice how the pairings all have negative free-energy values (more stability) and the loops have positive free energies (less stability) that vary depending on the size of the loop.

Comparing possible structures

The next step is to use the nearest-neighbor (Turner) free-energy values to determine the free energy of a given structure. Drawing all the different possible RNA structures for a given RNA sequence is beyond the scope of this book. However, it is important to point out that the number of possible RNA structures scales exponentially with the length of the sequence. In fact, an RNA sequence with 100 nucleotides has more than 10^{25} possible structures. Thus, the thermodynamic methods are restricted to working with shorter RNA sequences. Figure 5.1.2 shows the steps of calculating and comparing the free energies of two small RNA secondary structures for the same primary RNA sequence.

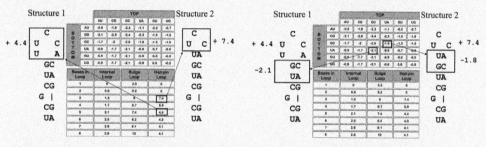

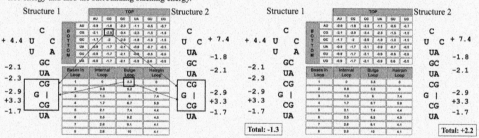

FIGURE 5.1.2. Calculating the total free energies of two possible secondary structures for the same RNA sequence: UCGCUGUUCCACAGGA. Structure 1 features a 5-nucleotide hairpin loop and a bulge. Structure 2 features a 3-nucleotide hairpin loop and a bulge. Structure 1 has the lowest free energy and, therefore, is the most stable of the two structures.

Algorithm 2: Mutual Information (MI)

The second method, known as MI, uses a comparative approach to RNA structure prediction. Specifically, it compares the variation in RNA sequences across related organisms and determines whether pairs of positions in the sequence alignment covary. In other words, when there is a mutational change at one sequence position, is there a corresponding change at another position? If this occurs multiple times at the same two positions, this is evidence that the positions are correlated (i.e., mutually informative) and that the positions are interacting in the molecule.

The following preparatory exercise should help you understand positional covariation in an RNA sequence alignment and how this evidence can be used to predict RNA structure.

Below is a sequence (sequence 1) of a known RNA structure:

Sequence 1: GAUCCUGCCUUCACGAUC

Here is the same sequence aligned with two other sequences:

Sequence 1: GAUCCUGCC--UUCACGAUC
Sequence 2: GACCCUGCC--UUCAGGGUC
Sequence 3: CAACCUGCCAGUUCACGUUG

The figure below shows the known RNA structure of sequence 1 on the left. The other sequences (2 and 3) have point mutations at positions 1, 3, 18, and 20. Fill in the different nucleotides for positions 3 and 18 in RNA sequence 2, and positions 1, 3, 18, and 20 in RNA sequence 3 in the spaces provided in the figure. Also, fill in the extra nucleotides for the 2-base AG nucleotide indel in sequence 3 at the top of the RNA structure. (Notice that the other sequences do not have that extra A and G, so the alignment is filled in with gaps.)

Sequence 1   Sequence 2   Sequence 3

Reflection
- How did the mutations affect the stem structure of the sequence 2 RNA? The sequence 3 RNA?
- How do positions 3 and 18 covary (i.e., how are they mutually informative)? How many times do they change together? How about positions 1 and 20?
- We know from the structure of sequence 1 that the position 2 nucleotide (A) is interacting with the second-to-last nucleotide (U) in the same sequence. Would MI help us predict this interaction? Why or why not?
- Notice that there seems to be an insertion of two extra nucleotides in the hairpin loop of sequence 3. The hyphens are put in the alignment to account for this indel mutation (insertion in sequence 3 or a deletion in the ancestor of sequences 1 and 2). How might this change the free energy of the structure?

The figure below shows the answer. Notice how the change in a nucleotide at one part of the stem in sequence 2 and sequence 3 (compared with sequence 1) correlates with a change at another part of the same sequence that maintains the base pair and the stem structure. This is a very common pattern in RNA sequence alignments—the most common, in fact. It makes sense because the stem structures are critical for the stability of the molecule, as you know from the previous section. Since the principle of MI relies on correlated changes (covariation) between nucleotide positions in a sequence alignment as evidence of interaction, this means two things: (i) if there are no changes, MI cannot predict interactions, and (ii) other correlated changes, such as pseudoknots or base triples, can also be detected.

| | Sequence 1 | Sequence 2 | Sequence 3 |

Figure 5.1.3 shows many instances of covariation in a larger alignment of 10 related RNA sequences and how to identify mutually informative nucleotide positions.

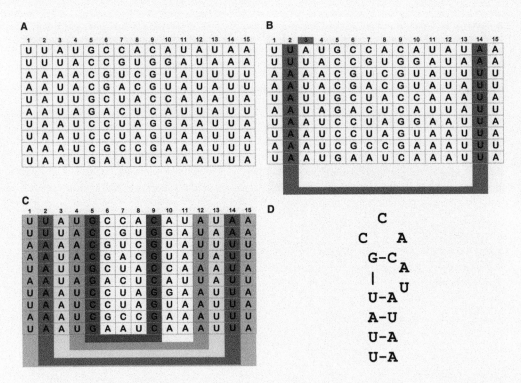

FIGURE 5.1.3. Principle of MI. This comparative approach requires a multiple-sequence alignment **(A)**. The method searches for correlated positions in the alignment. For instance, in this alignment, positions 2 and 14 appear to be correlated **(B)** because as position 2 changes, from an A to a U, for example, there is a corresponding change in position 14 from a U to an A. All instances of corresponding changes are identified **(C)**, and this can be used to determine the secondary structure of the RNA sequences. Finally, the predicted structure of the first sequence in the alignment is shown **(D)**. All the correlated changes in this example are base pairing in stem regions, but other changes are also possible to predict.

Exercises

Interactive exercises (theory)

Use the RNA free-energy and MI links below to learn how to determine the best RNA structure for a single sequence using thermodynamic calculations and to predict interacting positions using a multiple-sequence alignment and MI.

RNA Free-Energy Interactive Link
Link:
**http://kelleybioinfo.org/algorithms
/default.php?o=3**

Mutual Information Interactive Link
Link:
**http://kelleybioinfo.org/algorithms
/default.php?o=10**

Problems

1. Determine the RNA free energy of the following sequence. Show your work.

		TOP					
		AU	CG	GC	UA	GU	UG
B	AU	-0.9	-1.8	-2.3	-1.1	-0.5	-0.7
O	CG	-2.1	-2.9	-3.4	-2.3	-1.5	-1.5
T	GC	-1.7	-2	-2.9	-1.8	-1.3	-1.5
T	UA	-0.9	-1.7	-2.1	-0.9	-0.7	-0.5
O	GU	-0.9	-1.7	-2.1	-0.9	-0.5	-0.5
M	UG	-0.9	-1.7	-2.1	-0.9	0.6	-0.5

Bases in Loop	Internal Loop	Bulge Loop	Hairpin Loop
1	0	3.3	0
2	0.8	5.2	0
3	1.3	6	7.4
4	1.7	6.7	5.9
5	2.1	7.4	4.4
6	2.5	8.2	4.3
7	2.6	9.1	4.1
8	2.8	10	4.1

```
            A A
    U              G
     U           C
        G U
     C   |
        G U
        A U
     C        U
  A              C
  A           C
        G U
     G   |
        G U
     C   |
        U A
        U A
        A U
     C   |
     C   |
        C G
  5' U       C 3'
```

RNA Free Energy: ☐

☐

☐
☐
☐
☐

☐
☐
☐
☐
☐
☐

Total free energy: _____

2. a. Draw lines below the table connecting predicted interacting positions based on the principle of MI.

Length: 15

Number of Sequences: 10

RESET   CONCEPT MODE

1	2	3	4	5	6	7	8	9	10	11	12	13	14	15
A	G	U	C	G	A	A	C	C	C	G	A	C	U	G
U	G	U	C	C	A	A	A	U	G	G	A	C	A	G
U	G	U	C	C	G	U	C	C	G	G	A	C	A	G
A	C	A	C	C	A	A	U	A	G	G	U	G	U	C
U	C	A	G	G	C	A	A	G	C	C	U	G	A	G
U	C	A	C	C	G	A	G	C	G	G	U	G	A	G
A	G	A	G	C	G	C	G	C	G	C	U	C	U	A
U	C	U	G	C	U	U	C	A	G	C	A	G	A	C
A	G	A	G	C	C	A	G	G	G	C	U	C	U	G
U	G	A	G	C	A	C	G	U	G	C	U	C	A	C

b. Draw the predicted structure of the first sequence in the alignment (top row) below.

Lab Exercises (Practice)

In this part of the exercise, you will learn how to use programs for RNA folding and MI.

Mfold (Free Energy) Tutorial Link:
Link:
**http://kelleybioinfo.org/algorithms
/tutorial/TRna1.pdf**

MatrixPlot (Mutual Information) Tutorial Link:
Tutorial:
**http://kelleybioinfo.org/algorithms
/tutorial/TRna2.pdf**

Sample and lab exercise data (for both Mfold and MatrixPlot):
**http://kelleybioinfo.org/algorithms
/data/DRna1.txt**

Lab Exercise

Part 1. RNA folding

Use Mfold to predict the structure of a tRNA sequence and answer the following questions.

>Haemophilus_influenzae_tRNA
GGGGAUAUAGCUCAGUUGGGAGAGCGCUUGAAUGGCAUUCAAGAGGUCGUCGGUUC
GAUCCCGAUUAUCUCCACCA

1. What is *Haemophilus influenzae*?

2. What is a tRNA and what does it do?

3. Draw/show your tRNA sequence in the analysis window and answer the following questions or follow the directions below.

   a. Click on an energy dot plot file link. What is this energy plot telling you? Explain.

   b. Draw or paste the top left corner of the energy dot plot from positions 1 to 30. Your answer should include nine 10 × 10 position squares.

   c. Draw the RNA structural element predicted in this region of the dot plot. You should be able to find this element in the structure 1 image file.

   d. Compare the predicted RNA structure 1 to structure 2. How do they differ, and how might this change the free energy of the molecule? Explain.

Part 2

Use MatrixPlot to predict the structure of the FASTA-aligned sequences included with the sample data (RNA1 to RNA12).

http://kelleybioinfo.org/algorithms/data/DRna1.txt

1. Write the matching nucleotide positions of the longest predicted stem-loop region (the longest diagonal) in the graph. Approximate positions will suffice given the difficulty of reading the graph. (As the old saying goes, you get what you pay for, and these bioinformatics websites are free.)

2. Write the numbers of the predicted interacting positions indicated by the diagonal in the top left corner of the MatrixPlot graph. Note: there are 345 aligned sequence positions (*x* and *y* axes include positions 1 to 345).

3. Write the interacting RNA nucleotides at these positions using the RNA1 sequence.

Notes

1. Hydrogen bonding of guanine with uracil (considered noncanonical base pairs) is very common in RNA molecules.
2. **Turner DH, Mathews DH.** 2009. NNDB: the nearest neighbor parameter database for predicting stability of nucleic acid secondary structure. *Nucleic Acids Res* **38**:D280–D282.

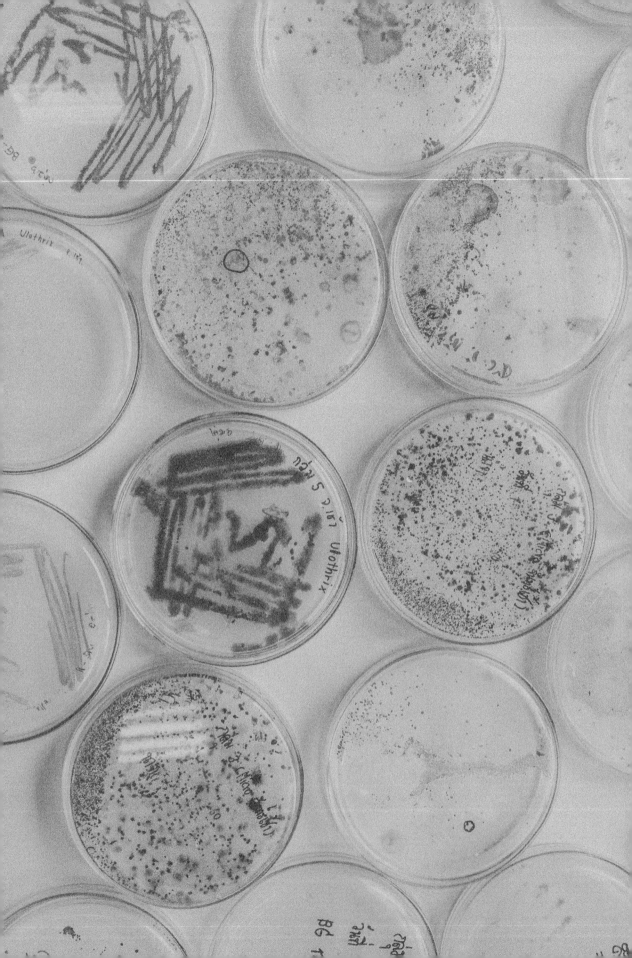

CHAPTER 06

PHYLOGENETICS

One of the most remarkable, and arguably the most important, bioinformatics discoveries revolutionized our entire understanding of life on Earth. In 1977, Carl Woese and colleagues at the University of Illinois manually aligned pieces of ribosomal RNA[1] sequences isolated from common bacteria (*Escherichia coli* and cyanobacteria), a few eukaryotes (yeast and an aquatic plant called duckweed), and some methane-generating "bacteria" previously isolated from dairy cows. Using some simple calculations to estimate how different the sequences were from one another and some parsimony-type reasoning (see Activity 6.1), the authors generated a small phylogenetic tree that completely upended our understanding of life on Earth (Fig. 6.1).

While the tree doesn't look like much, it strongly suggested that the methanogenic bacteria (e.g., *Methanobacterium*) and their relatives weren't bacteria at all. Rather, they comprised a major new branch in the tree of life. As you can imagine, this result caused quite a stir and was greeted with much skepticism. This new hypothesis of life overturned the basic 5-kingdom description of life that had been accepted since the mid-19th century. However, the more sequences scientists generated (including whole-genome sequences and alignments), and the better the alignment and phylogenetic algorithms, the clearer and more solid the pattern became (Fig. 6.2).

Ramifications of the "Big Tree"

This phylogeny not only changed our understanding of the evolution of life but also inspired the development of molecular techniques that allowed us to study microbes in basically any environment without having to culture them first. The DNA techniques invented by Norman Pace and colleagues (also at the University of Illinois) allowed researchers to find novel microbial life everywhere, including

- Boiling geothermal hot springs more acidic than battery acid
- Every rock, soil, and plant surface on the planet
- Deep inside mineshafts, below frozen lakes in Antarctica, and in the sediments at the bottom of the ocean

Phylogenetic Tree of Life

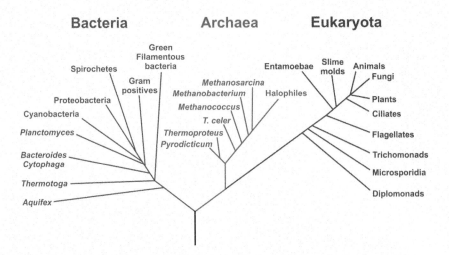

FIGURE 6.1. **The "universal" phylogenetic tree circa 1987.** The tree of life naturally separated into three domains: the *Eubacteria* (now called *Bacteria*), the *Eukaryota* (now *Eukarya*), and the then-newly identified group called the *Archaebacteria* because they were bacteria-like. The *Archaebacteria* turned out to be fundamentally different from the *Bacteria* and share many molecular and cellular aspects with the *Eukarya*, and they are now called the *Archaea*.

- In clouds, industrial waste pits, shower curtains, and scalding steam vents on Hawaiian volcanoes[2]
- The mouth, gut, and surface of every animal on Earth

The number of discoveries proceeding from this research has been mind-blowing. Since 1977 and the development of culture-independent molecular methods, we have learned that

- 99.999% of life is suspected to be microbial and has yet to be cultured[3]
- Bacteria account for the majority of the biomass on the planet
- Microbes exist that can survive boiling water, pH near 0, and salt concentrations greater than 25%
- Ocean life is mainly microbial, with every milliliter of seawater loaded with bacteria, archaea, eukaryotic cells, and even more viruses
- The number of species on Earth is estimated at greater than one billion[4]
- The number of microbial cells on planet Earth is ~10^{29} and the number of viruses is ~10^{31} [5]
- The human body hosts ~100 trillion microbes, 10 times more than human cells

We are now using these techniques, inspired by the simplest of sequence alignments and phylogenetics, to study microbes and their relationship to human health, discover new pathogens, track sources of pollution, study the oceans and the tundra, and discover new mechanisms to change the course of evolution.

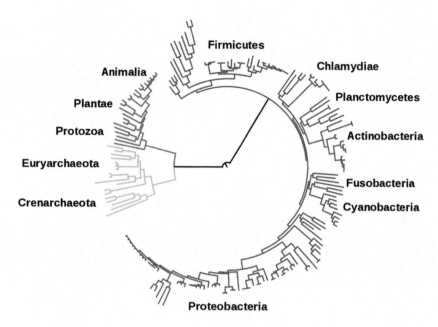

FIGURE 6.2.The expanding tree of life. New molecular methods, including whole genome sequencing and direct sequencing of DNA from complex environmental samples have greatly expanded our understanding of microbial diversity. The *Bacteria* are indicated in blue, the *Eukarya* in red, and the *Archaea* in green. This tree is biased towards bacterial lineages and only includes a few select representative sequences. However, it does illustrate how our understanding of microbial diversity has grown. A complete depiction of microbial diversity would include millions of branches and would be impossible to display in a figure. Still doubted by some in the early-2000s, the pattern of this phylogenetic tree continues to strengthen with the addition of genomic sequences.[6]

Uses of Phylogenetics

The use of phylogenetics long predates the work of Woese and colleagues, although the computational methods trace back only to the 1960s. Most phylogenetic trees were built to describe the relationships among macrofauna and -flora like plants, insects, birds, marmots, and whales. Prior to DNA sequencing, phylogenetic analysis relied on morphological data (presence or absence of scales, bones, and hair) and other visible characteristics. However, with the development of DNA sequencing methods, nucleotides and amino acids became predominant.

Beyond the big tree, phylogenetic theory has had an enormous impact on our understanding not only of the relationships among organisms but also of the relationships among genes within genomes, the process of molecular evolution, the evolution of gene families, patterns of recombination, and horizontal gene transfer. Phylogenies have also been used to discover new forms of life and the origins of deadly viruses and bacteria. Figure 6.3 shows some examples of the many possible uses of phylogenetics.

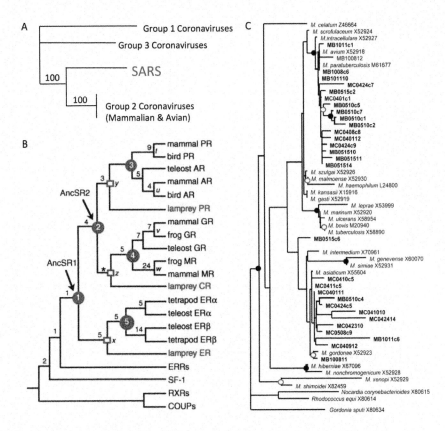

FIGURE 6.3. **Some uses of phylogenetic trees. (A)** Pathogen identification. Sequence alignment and phylogenetic analysis were used to show that the deadly severe acute respiratory syndrome (SARS) virus was a type of coronavirus (a common cold virus). The numbers indicated the maximum possible bootstrap values (see "The Bootstrap" later in this chapter). **(B)** Relationships among steroid receptors, transcription factors that bind steroid hormones like estrogen (ER, estrogen receptor) or testosterone (AR, androgen receptor). This phylogeny shows that all steroid receptors were derived from an ancestral protein that likely bound an estrogen-like hormone. PR, progesterone receptor; GR, glucocorticoid receptor; MR, mineralocorticoid receptor. The lamprey sequence outgroups are highlighted in red. Reprinted from **Thornton JW.** 2001. *Proc Natl Acad Sci U S A* **98:**5671–5676, with permission. **(C)** Phylogeny of cultured and uncultured mycobacterial species (*Mycobacterium tuberculosis* causes, you guessed it, tuberculosis). The boldface numbers indicate novel uncultured species of mycobacteria from the air of a hospital pool where the lifeguards had been getting sick and coughing up blood. Reprinted from **Angenent LT, Kelley ST, St Amand A, Pace NR, Hernandez MT.** 2005. *Proc Natl Acad Sci U S A* **102:**4860–4865, with permission.

How To Interpret Phylogenetic Trees

There are many ways to draw phylogenetic trees and different aspects of trees that require interpretation. A phylogeny is a hypothesis of the evolutionary relationships of organisms or genes usually based on molecular data. The minimum information a phylogenetic tree shows is the relationships among the taxa. You will see the word taxon (singular) or taxa (plural) used often with phylogenies. It comes from

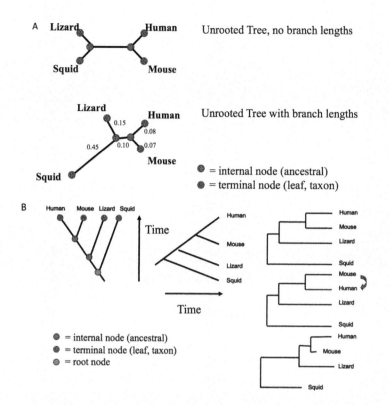

FIGURE 6.4. **Aspects of phylogenetic trees. (A)** The topology of the tree indicates the relationships among the taxa. The taxa at the tips (leaves, blue circles) of the tree are connected by branches to the other taxa via internal nodes (red circles). The nodes indicate the common ancestor, and the fewer the nodes between taxa, the closer their phylogenetic relationship. Both trees are unrooted; however, the bottom tree shows additional information in the form of branch lengths. Longer branches indicate more evolutionary change in the sequence since the split from the common ancestor. **(B)** On the left is a cladogram-type tree rooted with the squid (an invertebrate) outgroup. Outgroups are used to determine the order of evolutionary events. Squid make a good outgroup in this case because they are "outside" the group of three vertebrates. The trees to the right have the same topology, just sideways. The arrow shows that one can rotate taxa at a node without affecting the interpretation, since evolutionary time always extends from the root node outward.

the word taxonomy and is a general term to encompass any level of taxonomic organization. A taxon can refer to a species, genus, or family of organisms, but it can also refer to genes or even unknown groups. Figure 6.4 shows how to read different aspects of phylogenetic trees that can be used to describe the relationships among taxa, the amount of mutational change since the taxa split from a common ancestor, and even the statistical support of the relationships.

The Bootstrap

In Activity 6.1, we will cover two methods for creating phylogenetic trees using multiple-sequence alignments of DNA or protein sequences: distance and parsimony. These tree-building methods aim to determine the best phylogenetic tree for a given set of taxa. However, while these methods can build a phylogeny, they cannot by themselves determine the statistical significance of the tree.

This is where the bootstrap[7] method comes in. Phylogenetic bootstrapping is the most commonly used method for determining how well the data support the relationships in the tree. A bootstrap is a common statistical procedure that creates a random resampling of the data with replacement. In the case of phylogenetic analysis, the bootstrap resamples the positions of the multiple-sequence alignment, creating a new data set of the same size (same number of positions). In a typical bootstrap analysis, many hundreds or thousands of bootstrap replicates are performed, and each replicate creates a randomly sampled sequence alignment (Fig. 6.5A). During each replicate, a phylogenetic tree is built from the randomly as-

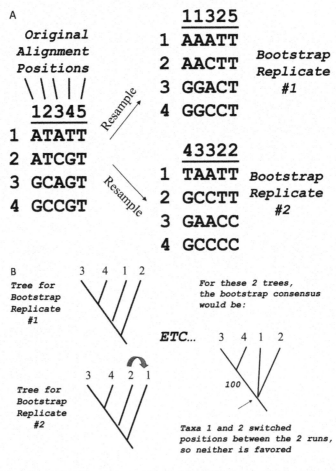

FIGURE 6.5. **Example of bootstrap analysis. (A)** Making two bootstrap data sets by sampling with replacement. One hundred bootstrap replicates would make 100 data sets. Notice how some positions have been sampled multiple times in a data set while others not at all. **(B)** A phylogenetic tree is built for each data set. All the resulting trees are combined to make a consensus tree. If all the trees have a particular node, this node is said to have 100% bootstrap support. In a standard approach, if a node is found in fewer than 50% of the bootstrap trees, all the unsupported branches are collapsed into a single node.

sembled alignment. If one performs 1,000 bootstrap replicates, one has created 1,000 phylogenetic trees of the same set of taxa. In the final step, all the bootstrap replicates are summarized into a single bootstrap consensus tree (Fig. 6.5B).

The idea behind the bootstrap in phylogenetics is simple: if every single tree in all the bootstrap replicates (100% of the trees) shows taxon A closely related to taxon B, this is the highest support that can be achieved and indicates that the relationship of taxon A to taxon B is strongly supported by the data. For example, in the SARS phylogeny (Fig. 6.3A), there is 100% support for the relationship of the SARS virus to the group 2 coronaviruses. In other words, no matter what positions of the alignment are selected, SARS is always closely related to the group 2 coronaviruses. Bootstrap values of 95% or higher are considered significant, though in practice values of 70% or higher can be trusted.

It is important to keep in mind several things about the bootstrap. First, all the bootstrap does is create randomized data sets. The bootstrap is a statistical method, not a phylogenetic method. The phylogenetic analysis is performed separately with each bootstrap data set. For instance, one can perform a neighbor-joining, a maximum-parsimony, or a maximum likelihood bootstrap analysis (as we will see in Activity 6.1). Second, the bootstrap phylogeny is a consensus phylogeny, not the best phylogeny for the data set. For instance, the best phylogeny may resolve all the relationships, but the bootstrap often collapses the branches that are not well supported (Fig. 6.5B). Third, the more bootstraps performed, the better. However, some methods are very slow (like maximum parsimony and especially maximum likelihood), and depending on the number of taxa, it may take too much time to do thousands of bootstrap replicates.

Notes

1. The sequences were actually RNAs isolated from organisms grown in cultures—a real pain in the neck! Now we just sequence the DNA that codes for the RNA (or anything else).
2. To read about those last two, see **Kelley ST, Theisen U, Angenent LT, St Amand A, Pace NR**. 2004. Molecular analysis of shower curtain biofilm microbes. *Appl Environ Microbiol* **70**:4187–4192 and **Ellis DG, Bizzoco RW, Kelley ST**. 2008. Halophilic Archaea determined from geothermal steam vent aerosols. *Environ Microbiol* **10**:1582–1590.
3. **Locey KJ, Lennon JT**. 2016. Scaling laws predict global microbial diversity. *Proc Natl Acad Sci U S A* **113**:5970–5975.
4. One study estimated total number of eukaryotic species at 8.7 million (**Mora C, Tittensor DP, Adl S, Simpson AG, Worm B**. 2011. How many species are there on Earth and in the ocean? *PLoS Biol* **9**:e1001127), while microbial diversity (mainly *Bacteria* and *Archaea*) has been estimated at 1 trillion (see endnote 7, above).
5. See **Kallmeyer J, Pockalny R, Adhikari RR, Smith DC, D'Hondt S**. 2012. Global distribution of microbial abundance and biomass in subseafloor sediment. *Proc Natl Acad Sci U S A* **109**:16213–16216 and **Whitman WB, Coleman DC, Wiebe WJ**. 1998. Prokaryotes: the unseen majority. *Proc Natl Acad Sci U S A* **95**:6578–6583.
6. Why only 3 domains—why not a 4th, 5th, or more? Have we missed them? Are viruses the 4th domain? Stay tuned!
7. The term "bootstrap" comes from the idiom "pull yourself up by your own bootstraps," which means to do it without any outside assistance (i.e., all by yourself). Bootstraps can be found on cowboy boots and are used to put them on. In the context of statistics or phylogenetics, bootstrap methods resample the same data as used to generate the result, and thereby to test the robustness of the result. "Pull the result up by its own data."

ACTIVITY 6.1   PHYLOGENETIC ANALYSIS

Motivation

A phylogeny is a diagram that indicates the evolutionary relationships among organisms or the molecular sequences (DNA, RNA, or protein) within organisms.[1] Phylogenetic trees have been used to, among other things, determine the relationships among songbirds, study the evolution of drug resistance in HIV, predict new gene functions, and discover new forms of microbial life. Phylogenetic methods, algorithms for computing phylogenies, require multiple-sequence alignments of DNA or protein sequences of the same gene from different organisms to determine the evolutionary relationships among the organisms.[2] One may also make sequence alignments of different but related genes (e.g., all gene sequences in the globin family) to determine how the gene family evolved. Phylogenetic methods use information from the mutations that have accumulated among the sequences over evolutionary time to determine how the sequences are related to one another.

In this activity, we will cover two basic methods for constructing the best phylogenetic tree, given a set of aligned sequence data. The underlying principles behind these methods are very different, but they have the same goal: finding the best phylogenetic tree for the data. We will also cover a statistical approach for assessing the quality of the phylogeny and use an online resource for building phylogenetic trees using these methods.

Learning Objectives

1. Learn the principles and goals of phylogenetics, some uses of phylogenies, how to interpret phylogenetic trees, and how the bootstrap method can be used to determine tree accuracy (Motivation).
2. Be able to calculate a distance matrix based on a multiple-sequence alignment and understand how it can be used to build a neighbor-joining tree (Concepts and Exercises).
3. Use the maximum-parsimony principle and set theory to determine the ancestral characters in a phylogenetic tree (Concepts and Exercises).
4. Use the maximum-parsimony principle to determine the best among a set of phylogenetic trees (Concepts and Exercises).
5. Learn how to use an online program for building phylogenetic trees using neighbor joining and maximum parsimony and run a bootstrap analysis (Concepts and Exercises).

Concepts

This activity covers two related, but fundamentally different, approaches to estimating the phylogenetic relationships among a set of aligned DNA or protein sequences. The first method we will cover is referred to as a distance method because it uses the overall distance (dissimilarity) among the sequences in a multiple-sequence alignment[3] to determine the relationships. The second method, referred to as maximum parsimony (MP), also uses a multiple-sequence alignment but focuses on specific nucleotides or amino acid positions in the alignment to determine how the sequences are related to one another. In science, the principle of parsimony states that in choosing between competing hypotheses, the one with the fewest assumptions should be selected.[4] In the case of phylogenetic trees, MP selects the phylogenetic tree that requires the fewest number of changes (i.e., the maximally parsimonious tree).

Algorithm 1: distance method
The idea behind distance-based methods is simple: the greater the similarity of two sequences (the shorter their distance), the more closely related they are to each other. Distance values range from 0 to 1, with sequences that are identical (all DNA nucleotides or protein amino acids) having a pairwise distance value of 0 and sequences that have nothing in common having a distance value of 1.

The following exercise should help you understand the principles behind distance calculations. Below is a multiple-sequence alignment of four DNA sequences, each from a different species of marmot. By comparing the sequences in the alignment, can you tell which species are more closely related? Which ones are distantly related?

```
Species  1  ATATTTCGAT
Species  2  ATCGTCCGGA
Species  3  GCCGTTCGCA
Species  4  GTAGTCGGAT
```

Closest (smallest distance) species:

Farthest (greatest distance) species:

Reflection
- What is the total number of identical nucleotides between species 1 and species 2? How about species 1 and species 3?
- Could you use the number of differences and the length (number of positions) in the sequence alignment to calculate a distance score between species 1 and 2?
- Could you use a similar approach with protein sequences? What would you compare?
- How many total pairwise comparisons are there in this alignment? If there were 100 sequences, would you offer your computer some candy if it helped you out?

The answer is as follows. The sequences for species 1 and species 4 have 4 out of 10 nucleotides that are different between them. The same is true of species 2 and species 3. Since the sequences are 10 nucleotides in length, the distance is 4 out of 10, or 0.4, for these pairs and makes them equally close matches. Species 1 and species 3, and species 3 and species 4, have 6 out of 10 nucleotides different, a distance of 0.6, which makes them equally far matches.

```
Species  1  ATATTTCGAT
Species  2  ATCGTCCGGA
Species  3  GCCGTTCGCA
Species  4  GTAGTCGGAT
```

Number of different nucleotides

1 — 2	5
1 — 3	6
1 — 4	4
2 — 3	4
2 — 4	5
3 — 4	6

Closest : 1 and 4, 2 and 3
Farthest : 1 and 3, 3 and 4

Using distances to build a phylogenetic tree

Distance methods estimate the phylogenetic relationships among a set of taxa by (i) determining all the pairwise distances between all sequences[5] in an alignment and then (ii) using these distances to build a phylogenetic tree. There are several ways to build phylogenetic trees based on a pairwise distance matrix. We will focus on a simple and widely used method called the neighbor-joining (NJ) algorithm. As its name implies, it builds a phylogenetic tree by "joining neighbors" sequentially, resulting in a single best tree for all the sequences. NJ is very fast, even with thousands of sequences, and it can be used with any distance matrix, which makes it the most popular algorithm for distance-based phylogenetic analyses.

Figure 6.1.1 details how to determine a distance matrix for a set of 4 related DNA sequences using a simple distance metric. Figure 6.1.2 describes a (simplified) iterative process of using a distance matrix to build an NJ phylogenetic tree.

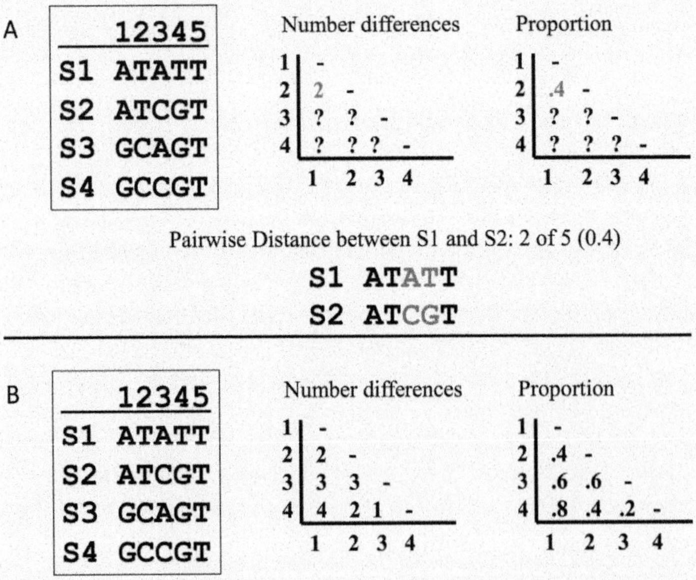

FIGURE 6.1.1. **Calculating a distance matrix with 4 aligned DNA sequences.**
(A) Distances are calculated between all pairs of sequences. Based on the sequence alignment in the upper left corner, sequences S1 and S2 differ at two nucleotide positions out of five total in the alignment, making the proportional difference 0.4. **(B)** All pairs of distances are calculated to complete the matrix.

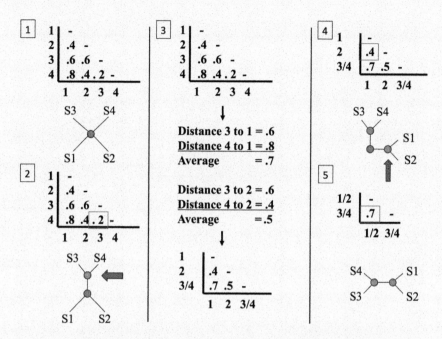

FIGURE 6.1.2. Using the distance matrix from Fig. 6.1.1 to build an NJ tree. (1) The initial tree is a so-called star phylogeny in which all the species are attached to a central node, and it has no structure. **(2)** The first step is to search the matrix for the nearest neighbors: the sequences with the shortest distance. These are joined together on the tree. **(3)** Next, the average distances are calculated for the two sequences that were joined (S3 and S4) to all the other sequences in the matrix. This creates a new, smaller distance matrix. **(4)** Repeat step 2 with the new distance matrix. Find the shortest distance and join the taxa. **(5)** Recalculate the distance matrix and join nodes until there are no more distances.

The online distance matrix interactive link (see Exercises) has additional explanations of how to build a matrix for a set of sequences and generate a phylogenetic tree. The distance calculations performed here are the simplest possible, and there are more sophisticated metrics that account for mutational biases and the possibility of mutation reversals. Also, the NJ tree-building approach is more complicated than described and includes estimations of branch lengths. However, the steps in Fig. 6.1.1 and 6.1.2 and the algorithm interactive show the basic principles of matrix building and NJ tree construction.

Algorithm 2: maximum parsimony (MP) method

MP uses the same data (a multiple-sequence alignment) as distance methods, but it doesn't create a matrix of similarities. Instead, MP examines each position in a multiple-sequence alignment separately to see if it has any useful information. If so, MP determines what each of the informative positions says about how the sequences are related.

To gain a sense of how this works, try the following problem. In this multiple-sequence alignment of protein sequences from 4 different species of marmots, try to identify amino acids that are shared between species. For example, species 1 and species 4 share a Y at position 2, while species 2 and 3 share an A at the same position. Write down which positions are shared between the sequences below.

```
          1 2 3 4 5 6 7 8 9 10
Species 1 V Y L G H E F Q K S
Species 2 V A L R H D F Q K W
Species 3 V A L R H D F Q F W
Species 4 V Y L G H E F Q F S
```

Positions with shared amino acids

Species 1 and 2:
Species 1 and 3:
Species 1 and 4:
Species 2 and 3:
Species 2 and 4:
Species 3 and 4:

Reflection

- Based on your positional analysis of this multiple-sequence alignment, can you conclude what species are most closely related?
- What do positions 1 and 3 tell you about which species are more closely related?
- Are there any conflicting data? In other words, do the shared amino acids at some positions contradict the patterns at other positions?
- Could a similar approach be used with DNA sequences?

Here are the answers. Positions 2, 4, 6, and 10 indicate a close relationship between species 1 and 4 and species 2 and 3. Position 9 contradicts this result, and the rest of the positions do not say anything about which sequence is related to which according to MP.

```
          1 2 3 4 5 6 7 8 9 10
Species 1 V Y L G H E F Q K S
Species 2 V A L R H D F Q K W
Species 3 V A L R H D F Q F W
Species 4 V Y L G H E F Q F S
```

Positions with shared amino acids

Species 1 and 2:	135789		
Species 1 and 3:	13578		
Species 1 and 4:	13578246	and	10
Species 2 and 3:	13578246	and	10
Species 2 and 4:	13578		
Species 3 and 4:	135789		

The basic principle behind MP is that only shared derived positions (called synapomorphies)[6] are useful for determining phylogenetic relationships. In the example, this includes positions 2, 4, 6, 9, and 10. Positions that are shared in all species are useless for MP, because they don't give you any information about whether species 1 is closer to species 2 than it is to species 3 or species 4. These positions are considered to be ancestral character states (derived from a single common ancestor species), and so the algorithm ignores them.

Which phylogenetic tree is the shortest (most parsimonious)?

Unlike the NJ distance method, MP does not build a single best tree. Instead, it chooses the best, most (maximally) parsimonious tree among a set of trees. Figures 6.1.3 to 6.1.6 show how MP determines which of two possible trees is the most parsimonious. The algorithm actually has two distinct but related aspects. The first part of the process uses principles of set theory to reduce the number of possibilities at each node of the tree. The second part finds the minimum number of changes that could have occurred to produce the pattern of nucleotides.

Figure 6.1.3 provides a problem with a small sequence alignment and two phylogenetic trees that show the possible relationships among 5 sequences (from 5 different taxa). There are actually 105 possible trees with 5 different taxa, but 2 are sufficient to show the basic process.

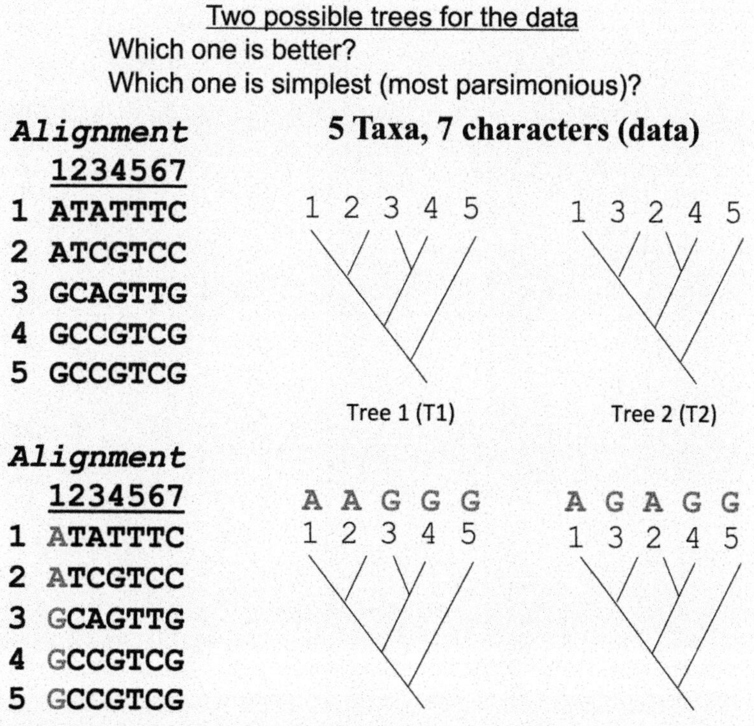

FIGURE 6.1.3. **Two possible phylogenetic trees for one data set of 5 sequences with 7 nucleotide alignment positions.** The bottom half of the figure shows the nucleotides of the first position mapped to the tips of the phylogenetic tree. Next, we will determine the minimum number of changes needed on the phylogeny to produce this pattern of nucleotides at this position (see Fig. 6.1.4 and 6.1.5).

In Fig. 6.1.4, set theory[7] is used to determine the most parsimonious set of nucleotides[8] at each node of the tree. The following describes the set theory notation for solving this part of the problem. (You will likely find the concept easier to grasp in the figure and the online tutorial/interactive.) The set theory logic is as follows:

{node} = {higher node 1} ∩ {higher node 2}

Translated into English, if the two higher nodes have something in common, the node contains only what they have in common (the intersection, ∩). For example:

{node} = {A, G} ∩ {G}

{node} = {G}

But what happens if they do not have anything in common? Easy, you keep everything! This is also known as the union (∪) of the two sets.

If {higher node 1} ∩ {higher node 2} = ∅

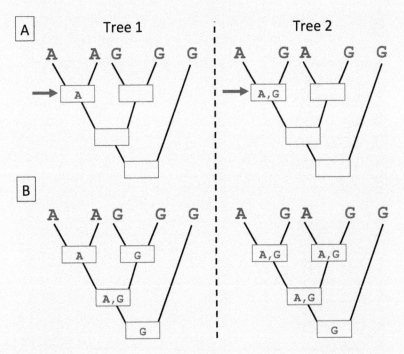

FIGURE 6.1.4. Using principles of set theory to reduce the number of possibilities at each node. To solve the problem, start at the tips (leaves) of the tree and move towards the base (root). **(A)** In tree 1, the first node is an A because the intersection of the two tips above is A. In tree 2, the first node is either an A or a G. The intersection of {A} and {G} is the empty set, so we keep everything—the union {A, G}. **(B)** Proceed down to the base of each tree using the above tips/nodes to determine the most parsimonious possibilities at each node.

{node} = {higher node 1} ∪ {higher node 2}

For example:

If {higher node 1} ∩ {higher node 2} = {G} ∩ {A} = ∅

{node} = {higher node 1} ∪ {higher node 2} = {G} ∪ {A} = {G, A}

Figure 6.1.5 shows the second part of the process: determining a path through the tree that requires the least number of changes (i.e., is maximally parsimonious). The online tutorial allows you to practice with this concept of minimization.

Finally, the minimization process is repeated for each position in the sequence alignment (Fig. 6.1.6). While tedious, it becomes quicker when one realizes that certain positions can be ignored because they are parsimoniously uninformative.

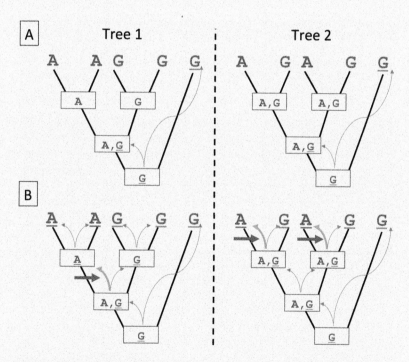

FIGURE 6.1.5. Having already minimized the possibilities at each node, choose the characters (nucleotides) at each node that require the fewest changes. (A) Starting at the root of the tree, the only option in both trees is G. Moving up the rightmost branch, we see that no changes are required. The position doesn't mutate (change) along the branch, and 0 steps are required. Moving up the left branch, if we choose G at the node, no changes are required. Choosing A would require 1 mutation (step) from a G to an A. This is not the minimum, so choosing G is preferable for both trees. **(B)** Complete the minimization for all nodes. In tree 1, the path chosen requires 1 total change in the tree for this position, the G-to-A mutation indicated by the arrow. No other path will be less than 1 for this position in this tree. In tree 2, the path we choose has 2 mutations, and this is also the minimum for this position on this tree.

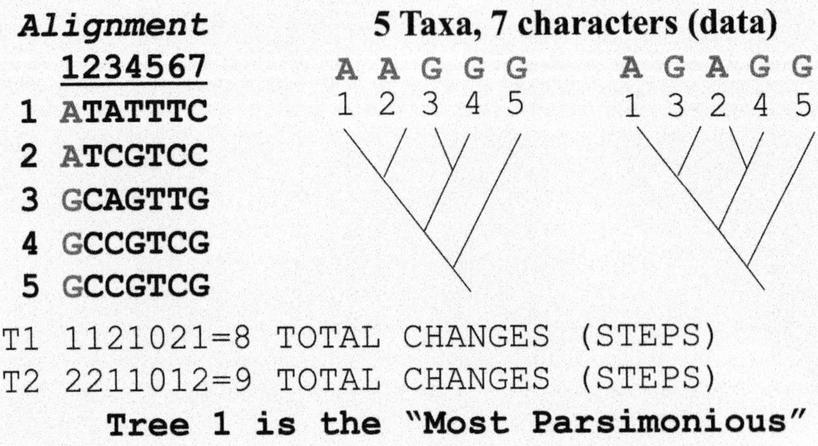

FIGURE 6.1.6. **Finish the problem by calculating the number of steps for each position on each tree.** Then, add them up to determine the total number of steps across all positions for each tree. In this case, tree 1 has a minimum total of 8 steps across all positions in the alignment. This is one step shorter than is needed for tree 2, so tree 1 is the most parsimonious option.

For example, if all the nucleotides are the same at a position, there will be 0 changes (steps) on all the possible trees at this position (see position 5 in Fig. 6.1.6). Similarly, if there has been only 1 change in one of the sequences, and all the rest are the same, the number of changes (steps) for this position will always be 1 no matter which tree is tested (position 4 in Fig. 6.1.6). These positions are called uninformative characters because they cannot inform which of the trees are most parsimonious. In fact, MP tree-building software completely ignores these positions.

The online MP tutorial explains how to determine the best phylogenetic tree given a multiple-sequence alignment:

http://kelleybioinfo.org/algorithms/interactive/lMax.pdf.

Exercises

Interactive exercises (theory)

Use the online phylogeny interactive links below to learn how to build distance matrices and phylogenies using the distance method and learn how to select the better phylogeny under the MP criterion. Once you learn how they work, solve the activity problems.

Distance Matrix Interactive Link
Link:
http://kelleybioinfo.org/algorithms /default.php?o=5

Maximum Parsimony Interactive Link
Link:
http://kelleybioinfo.org/algorithms /default.php?o=12

Problems

1. Distance matrix problem: answer the questions below.

    a. Fill in the tables.

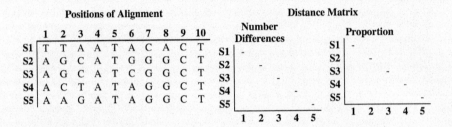

    b. Determine the first node of the tree and write it out in set notation (see the interactive).

FIRST TREE NODE:

    c. Recalculate the distance matrix proportions after joining the first node.

RECALCULATED DISTANCE MATRIX:

2. Parsimony problem: fill out the nodes in the tree below using parsimony. Calculate the length of the tree for this position and write it next to each tree.

	1	2	3	4	5	6	7	8	9	10
S1	R	A	I	I	F	N	P	W	M	E
S2	E	H	F	Q	F	U	P	W	M	E
S3	R	A	I	I	F	N	A	H	M	S
S4	E	H	F	Q	F	U	A	H	M	E
S5	R	Y	I	C	P	U	A	H	M	E

Tree 1:

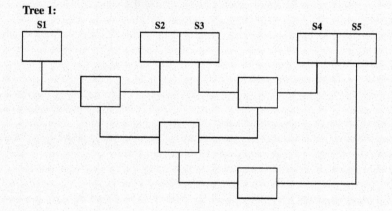

Tree 2:

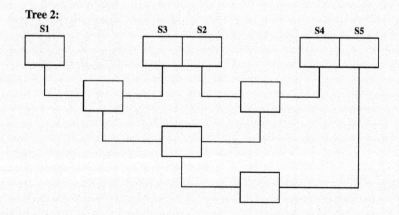

Lab Exercises (Practice)

In this part of the exercise, you will learn how to use a phylogenetic analysis website.

Phylogenetic Analysis Tutorial Link
Link:
**http://kelleybioinfo.org/algorithms
/tutorial/TPhy1.pdf**

Sample and lab exercise data (sample data 1):
**http://kelleybioinfo.org/algorithms
/data/DPhy1.txt**

Sample and lab exercise data (sample data 2):
**http://kelleybioinfo.org/algorithms
/data/DPhy2.txt**

Lab Exercise

Using the DNA and protein multiple-sequence alignments (sample data 1 and sample data 2, respectively), perform the phylogenetic analyses using the online phylogenetic software described in the tutorial.

1. Using the sample data specified below, draw/show the following results.

   NOTES:
   - In the Workflow Settings, uncheck the boxes for *Multiple Alignment* and *Alignment curation.* The sequences are already aligned and curated. Also, leave the *Visualization* box checked and choose *TreeDyn* if not already checked, and the workflow should be run "all at once."
   - Choose protein or DNA/RNA on the Input Data page as appropriate.
   - Leave the rest of the default settings on the Input Data page.

   a. Sample data 1, NJ tree (FastDist + Neighbor). Draw/show the tree below.

   b. Sample data 1, parsimony (TNT). Draw/show the tree below.

2. Analyze sample data 2 using the NJ approach (FastDist + Neighbor). Draw the top four branches of the tree or paste the tree below.

3. Perform a bootstrap analysis with 100 resamplings of the data using sample data 1 with the NJ criterion selected (FastDist + Neighbor). Draw/show the results below. Circle the branch with the highest bootstrap support.

Notes

1. DNA and other molecules are most often used to determine the relationships among organisms. However, sometimes researchers are interested in the evolution of the molecules themselves (e.g., gene family expansion, the evolution of drug resistance in HIV, and RNA structural changes).

2. Information from multiple-sequence alignments of different gene sequences within organisms can also be combined to increase information and phylogenetic accuracy.

3. See multiple-sequence alignment in Chapter 03, Activity 3.1.

4. Parsimony is also known as Occam's (or Ockham's) razor, named after William of Ockham, an English Franciscan Friar (c. 1287–1347) from the town of Ockham. The town is also known for opening its doors to William and Ellen Craft, slaves who escaped the United States after the passage of the barbaric Fugitive Slave Act of 1850 and who became important figures in the abolitionist movement. And it has a cool mill. They are very proud.

5. Distances can be calculated between pairs of aligned DNA/RNA sequences, protein sequences, and even organism characteristics like number of toes, hair color, behaviors, and other things.

6. Synapomorphy is derived from the Greek: "syn-" means "shared," and "-morph" means "shape."

7. In case you've forgotten (or never knew) how sets work, here is a short primer. The intersection of two sets is symbolized by $\cap$ and is equal to the set of what is shared between the two sets. The union of two sets is symbolized by $\cup$ and is equal to the set of all objects in both sets. For example, if $A = \{1, 3\}$ (set A includes the number 1 and 3) and if $B = \{3, 4\}$, then $A \cap B = \{3\}$ because the number 3 is all they have in common, and $A \cup B = \{1, 3, 4\}$, which is all the numbers in both sets. And finally, there is the concept of the empty set, symbolized by $\{\}$ or $\varnothing$, a set with nothing inside it. For example, if $A = \{1, 2\}$ and $B = \{3, 4\}$, then $A \cap B = \varnothing$ (the empty set) because the two sets share nothing in common.

8. This example uses DNA nucleotides, but the data could easily be amino acids or even physical (morphological) characteristics of organisms. Phylogeneticists refer to these generally as characters and the number of changes as steps.

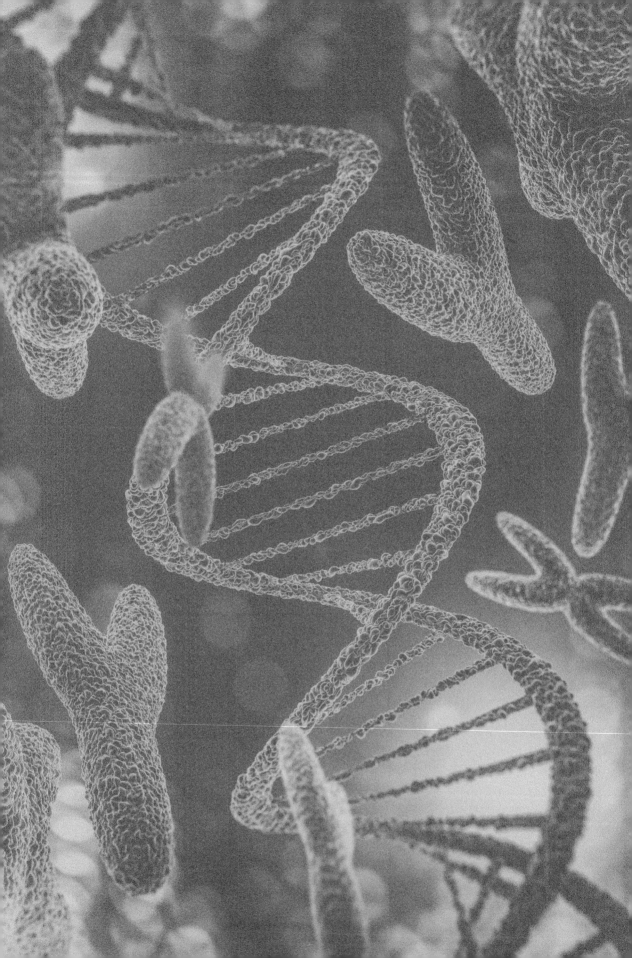

PROBABILITY: ALL MUTATIONS ARE NOT EQUAL (-LY PROBABLE)

You have already encountered the use of probability in several algorithms in this book, including the Chou-Fasman propensities in Chapter 02 as well as the sequence logos and position-specific weight matrices in Chapter 04. In this chapter, we discuss the concepts behind and generation of probability matrices that are used in many bioinformatics algorithms. Most of the chapter and exercises focus on how to determine the probability (likelihood[1]) of amino acids mutating to other amino acids. These amino acid substitution matrices—Point Accepted Mutation (PAM) and BLOcks SUbstitution Matrix (BLOSUM)—dramatically improve the performance of different protein alignment algorithms. We also briefly discuss the advanced concept of hidden Markov models (HMMs), a powerful means for making probability matrices "on the fly." HMMs are used in many bioinformatics applications, including predicting genomic repeat regions, transmembrane proteins, and protein-coding regions in genomes and clustering distantly related protein sequences into families.

Protein (Amino Acid) Substitution Matrices

One early realization made while analyzing protein multiple-sequence alignments was that not all amino acid mutations occur with the same frequency. You should have familiarity with amino acid substitution matrices from working with them in the protein Needleman-Wunsch analysis in Activity 3.1 (see PAM250 and BLOSUM62 links in Fig. 7.1).

Instead of using fixed match and mismatch values such as those used in DNA sequence alignments, alignments of protein sequences use a matrix of log odds[2] scores for matches and mismatches. The radio button on the website's interactive module is set by default to the PAM250 matrix, but one can also use the BLOSUM62 matrix. Both PAM and BLOSUM matrices contain log odds scores for every amino acid changing to every other amino acid, as well as not changing (Fig. 7.2). Note that the substitution matrices are symmetrical. For example, the log odds score of a change of arginine (R) to serine (S) is the same as for S to R.

Sequence Alignment

| MAIN PAGE |
| PRINT SCREEN |

Methods

BLAST Exercise **I**

Needleman-Wunsch **I**

Web Resources

NCBI BLAST **T** **D**

Clustal Omega **T** **D**

PAM 250 : ⊙

BLOSUM 62 : ○

Gap : -5

Sequence 1: GIVPWVK

Sequence 2: GIVWVK

SUBMIT

FIGURE 7.1. Screenshot of the Needleman-Wunsch interactive module for teaching protein sequence alignments using amino acid substitution matrices. By default, the traceback calculations use the PAM250 substitution matrix.

The BLOSUM matrices also contain log odds scores, but they are calculated in a different way, as you will learn later in the chapter.

These substitution matrices have proven very helpful for improving the performance of many algorithms, particularly sequence alignment methods such as dynamic programming and BLAST. The reason they are so useful in sequence alignment is that the scores really help differentiate among possible alignments. For instance, below are two small sequence alignments of the same query (QY) sequence to two different subject sequences (S1 and S2).

QUERY SEQ: HLRWS

SUBJECT 1 : HLSES

SUBJECT 2 : MLSWS

Alignment 1

| QY: | H | L | R | W | S |
| S1: | H | L | S | E | S |

Alignment 2

| QY: | H | L | R | W | S |
| S2: | M | L | S | W | S |

Using the PAM250 matrix in Fig. 7.2, one can score both by adding up the log-odds scores for all the alignment positions. For instance, in alignment 2 the score for an H (histidine)-to-M (methionine) change is –2, that for an L-to-L change is +6, and so forth.

In this case, even though both alignments have two mismatches, the PAM scoring system tells us that alignment 2 is a better alignment because it has a higher overall score.

Alignment 1

QY:	H	L	R	W	S	
	+6	+6	+0	−7	+3	=+8
S1:	H	L	S	E	S	

Alignment 2

QY:	H	L	R	W	S	
	−2	+6	+0	+17	+3	=+24
S1:	M	L	S	W	S	

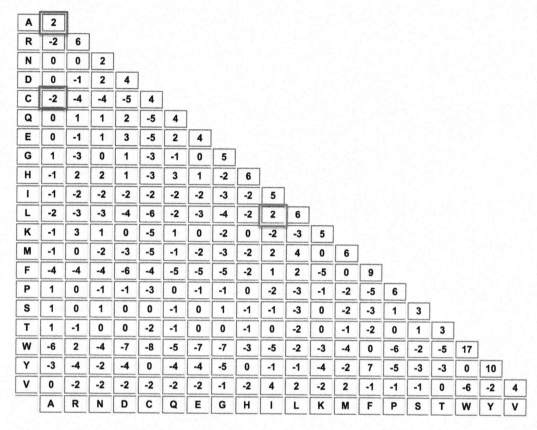

FIGURE 7.2. The PAM250 substitution matrix. This matrix shows the log odds scores of every amino acid changing to every other amino acid, or not changing at all. Positive numbers mean that the likelihood is higher than expected by chance, 0 means the same as chance, and negative means less likely than chance. For example, alanine not mutating (A to A on the table, top left red box) is more likely (+2) than alanine mutating to cysteine (A to C, −2, fifth-row red box). All the values along the diagonal are for the amino acid NOT changing. Typically, the log odds of no change are the highest values, but some substitutions are very common. For instance, a change from an isoleucine to a leucine (+2) is just as likely as an alanine not mutating.

This dramatic difference between the alignment scores (remember this is a log scale) is because W (tryptophan) has a very high score (+17) for not changing. In alignment 1, a change from E (glutamic acid) to W is very unlikely (−7) according to the PAM250 matrix. This is a 24-point difference on a logarithmic scale for the alignment of just one amino acid (tryptophan is weird). It should be noted that while the BLOSUM62 matrix has different values, the trend is still the same (i.e., W to W is +11) even though the method for determining the BLOSUM values is significantly different (see Activity 7.1).

What Determines Substitution Bias?

In the PAM250 substitution matrix, there is a high likelihood of an isoleucine (I)-to-leucine (L) change and vice versa (+2; Fig. 7.2). On the other hand, there is a low

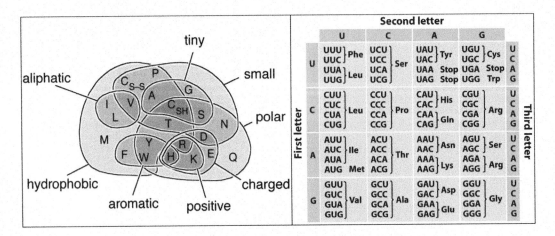

FIGURE 7.3. Venn diagram of amino acid properties (left) and the genetic code (right). Amino acids within circles of the Venn diagram are considered more biochemically similar.

likelihood (–3) of isoleucine being changed to glycine (G). Why is this? There are two fundamental reasons, which are illustrated in Fig. 7.3. First, isoleucine and leucine are "chemical cousins." These amino acids are similarly sized and have similar biochemical properties; namely, they are both hydrophobic (Fig. 7.3, left side, aliphatic grouping). A change of I to L in any given protein will usually have a very modest effect on the protein's function. However, isoleucine and glycine are very different biochemically, and a substitution of one for the other could spell disaster. See how in Fig. 7.3, left side, I and G are not in any shared groupings? If such a substitution were to occur in a critical cellular protein, it could eliminate an organism's ability to function or reproduce (Darwin says, "Goodbye!").

Second, the likelihood of a substitution is also dependent upon how many nucleotide changes (mutations) need to occur in the underlying protein-coding DNA. For example, the genetic code table shows that only one nucleotide in the first position of the codon needs to change (AUU to CUU) to cause an isoleucine-to-leucine mutation (Fig. 7.3, right side). On the other hand, an isoleucine-to-glycine mutation requires two independent mutations in the codon (AUU to GGU). Since the probability that two independent events will occur is the product of each independent probability,[3] the differences in codon sequences have a considerable impact on the likelihood of amino acid mutations.

Creating a probability model that takes into account both the biochemical properties of the amino acid AND the likelihood of mutations at the DNA level would be difficult, to say the least. Instead of doing this, a researcher named Margaret Dayhoff tried another approach: counting the different amino acid substitutions that occur in nature.

PAM and BLOSUM

Dayhoff's PAM amino acid substitution matrix and Henikoff and Henikoff's BLOSUM substitution matrix both estimate rates of substitutions using protein multiple-sequence alignments. Rates of substitution are estimated by counting the times amino acids change to other amino acids and asking whether this is more likely or less likely than expected by chance. Finally, the ratio of these values is used to determine the log odds scores. Activity 7.1 explains the differences between PAM and BLOSUM and the steps of calculating these matrices.

Multiple different substitution tables have been created using both the PAM and BLOSUM approaches. The choice of using a PAM or BLOSUM substitution matrix for BLAST or other applications should ideally be based on the overall dissimilarity of the sequences being compared or aligned. The greater the dissimilarity, the higher the PAM substitution matrix number that should be used. A PAM1 matrix might be more appropriate for very closely related sequences, while a PAM250 matrix may be more appropriate for highly divergent sequences. The numbers indicate the expected rate of mutation among sequences per 100 amino acids.

- PAM1: 1% mutation rate (1 mutation per 100 amino acid positions)
- PAM50: 50% mutation rate (50 mutations per 100 positions)
- PAM250: 250 mutations per 100 positions (positions mutate more than one time)

Like PAM, BLOSUM numbers also indicate divergence levels of the sequences used to make the matrices, but the values go in the opposite direction: lower numbers are used for more divergent sequences, higher numbers for less divergent sequences. A BLOSUM62 matrix was generated from sequences with an average of 62% overall similarity, while a BLOSUM90 matrix was generated from sequences that were 90% similar. Similarity equivalences between the two matrices are as follows:

PAM	BLOSUM
PAM100	BLOSUM90
PAM120	BLOSUM80
PAM160	BLOSUM60
PAM200	BLOSUM52
PAM250	BLOSUM45

When choosing what matrix to use, it is best to pick the matrix that fits the general divergence of the sequences being aligned. In practice, the choice of matrix does not make a particularly big difference in alignments.

Hidden Markov Models

The generalized probabilistic approach known as HMMs has been broadly applied in bioinformatics. Outside the fields of engineering and computer sciences, HMMs found application in the field of speech recognition in the 1970s. Shortly thereafter, early bioinformaticians applied them to the analysis of biological sequence data, especially protein sequences, though they have also found a number of other applications. While the mathematics behind HMMs are beyond the scope of this book, it is important to have an appreciation for HMMs and how they are used in bioinformatics.

In Activity 2.1 you gained some familiarity with the TMHMM server, an HMM used to predict the location of transmembrane domains in protein sequences.

HMMs have been used in bioinformatics to predict novel genes in genomes, perform multiple-sequence alignments, and fold RNA structures.[4] The genius of HMMs is their ability to find a sequential pattern given a sufficient set of known examples. For instance, protein sequences of known transmembrane domains provide the input for creating an HMM for detecting transmembrane domains in protein sequences of unknown function.

HMM can be defined as "a statistical Markov model in which the system being modeled is assumed to be a Markov chain with unobserved (hidden) states."[5] Nice. But what is a Markov chain, and what is meant by "hidden states"? A Markov chain is a statistical model that says that the probability of the next item in a sequence depends only on the state of the previous item in the sequence. For instance, one could have a Markov chain for predicting the weather. In a Markov chain, the chance it will be sunny tomorrow would depend on the weather the previous day (rainy or sunny). In the transmembrane domain example, this might mean that the probability of an amino acid in a sequence (Y, for example) being part of a transmembrane domain is dependent on the previous amino acid in the sequence (M, for example). In a Markov chain, one uses a transition matrix which has the probabilities of all the possible transitions from one state to another, much like the PAM and BLOSUM matrices.

If one knows all the transition probabilities for, say, the amino acids in a transmembrane domain, the problem is easy. However, since these probabilities are unknown they are considered hidden probabilities, and these hidden transitions must be inferred from real existing data. To determine an HMM, one must have a lot of known data of a particular type. This could include known transmembrane domain sequences, groups of intron sequences (for intron prediction), various alpha helices, etc. The HMM procedures "train" on specific data sets and produce an HMM specific for detecting new variants of the same type. In the lab activity, you will be using the Pfam HMM-based database to group a mystery sequence into a protein family.

Notes

1. In common speech, probability and likelihood are used interchangeably, though there are some differences—at least to statistics nerds. Both terms are used in this chapter, though likelihood is the more appropriate term for the chance that one amino acid mutates into another amino acid.
2. The log odds score is the logarithm of the odds ratio. In the case of amino acid substitution matrices, the odds ratio is the ratio of the observed likelihood of a substitution to the likelihood expected by chance.
3. It has been estimated that the chance of you getting caught in a tornado is 1×10^{-6}, while you have a 1×10^{-7} chance of being bitten by a shark. So, the expected probability of you being in a tornado AND being bitten by a shark is $(1 \times 10^{-6}) \times (1 \times 10^{-7}) = 1 \times 10^{-13}$. Unless you get caught in a sharknado. Then the probability increases to 1×10^{-1}.
4. For a review, see **Yoon B-J.** 2009. Hidden Markov models and their applications in biological sequence analysis. *Curr Genomics* **10**:402–415.
5. **Black EF, Marini L, Vaidya A, Berman D, Willman M, Salomon D, Bartholomew A, Kenyon N, McHenry K.** 2014. Using hidden Markov models to determine changes in subject data over time, studying the immunoregulatory effect of mesenchymal stem cells. *Proc IEEE Int Conf Escience* **1**:83–91.

ACTIVITY 7.1   GENERATING PAM AND BLOSUM SUBSTITUTION MATRICES

Motivation

Protein sequence matching (e.g., BLAST), multiple-sequence alignments, and phylogenetic analyses have long used substitution matrices to determine the best alignments or the best trees. Substitution matrices provide scores, called log odds scores, indicating the likelihood of each of the 20 most commonly occurring amino acids mutating into all the other amino acids or not mutating at all. The two approaches taught in this chapter for creating substitution matrices include PAM (Point Accepted Mutation) and BLOSUM (BLOcks SUbstitution Matrix). Both approaches use observed patterns of amino acid substitutions to generate substitution matrices. PAM determines these probabilities using phylogenetic trees, while BLOSUM bases the probabilities on conserved blocks of aligned sequences. Both methods also calculate how often the mutations are expected to occur by chance based on the frequency of the amino acids. The scores are logs of the ratios of the observed probabilities to the expected probabilities.

This activity will teach the principles of two different approaches for generating amino acid substitution matrices, as well as how to calculate the log odds scores. The lab exercises will show how these methods are used in protein BLAST analyses and introduce the Pfam (protein family) database, which uses the more sophisticated HMM approach to cluster groups of functionally related proteins.

Learning Objectives

1. Learn the biological principles behind substitution matrices and how probabilistic approaches can be used in bioinformatics (Motivation).
2. Use phylogenetic trees to estimate amino acid substitution rates and generate PAM-like substitution matrices (Concepts and Exercises).
3. Use conserved blocks within protein sequence alignments to estimate amino acid substitution rates and generate BLOSUM-like substitution matrices (Concepts and Exercises).
4. Learn how PAM and BLOSUM scores are used in protein BLAST analysis (Concepts and Exercises).
5. Gain familiarity with the HMM-based Pfam database (Concepts and Exercises).

Concepts

To better understand the biological principles behind amino acid substitution matrices, try the following exercise. On the next page is a Venn diagram illustrating the biochemical similarities among the 20 most common amino acids. The curved lines encompass letter designations for amino acids with similar biochemical properties, which are labeled outside the diagram. For example, isoleucine (I), valine (V), and leucine (L) are all aliphatic amino acids that also belong to a larger set of hydrophobic amino acids.

The protein sequence alignments below are of the same length and have the same number of identities (4 identical matches out of 10, i.e., 4/10). Use the information in the Venn diagram to determine which of the two matches is biochemically more likely.

Match 1

Query: **F G Q V I P A K R**

Subjt: **F A N V M P A R E**

Match 2

Query: **F G Q V I P A K R**

Subjt: **F S C V F P A F V**

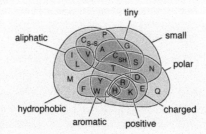

Reflection

- Are some mismatches more useful than others for determining the better match? Why or why not?
- Based on the Venn diagram and your understanding of proteins, would an E-to-M mutation be more or less likely than an E-to-N mutation? Why?
- Can you use the PAM log odds matrix below to score the two alignments? The scores are additive and the more positive, the more likely the substitution (or no substitution). For example, a change from N to N is +2, while a change from N to Y or Y to N is –2.

Match 1

Query: **F G Q V I P A K R**

Subjt: **F A N V M P A R E**

SCORE:

Match 2

Query: **F G Q V I P A K R**

Subjt: **F S C V F P A F V**

SCORE:

	A	R	N	D	C	Q	E	G	H	I	L	K	M	F	P	S	T	W	Y	V
A	2																			
R	-2	6																		
N	0	0	2																	
D	0	-1	2	4																
C	-2	-4	-4	-5	4															
Q	0	1	1	2	-5	4														
E	0	-1	1	3	-5	2	4													
G	1	-3	0	1	-3	-1	0	5												
H	-1	2	2	1	-3	3	1	-2	6											
I	-1	-2	-2	-2	-2	-2	-2	-3	-2	5										
L	-2	-3	-3	-4	-6	-2	-3	-4	-2	2	6									
K	-1	3	1	0	-5	1	0	-2	0	-2	-3	5								
M	-1	0	-2	-3	-5	-1	-2	-3	-2	2	4	0	6							
F	-4	-4	-4	-6	-4	-5	-5	-5	-2	1	2	-5	0	9						
P	1	0	-1	-1	-3	0	-1	-1	0	-2	-3	-1	-2	-5	6					
S	1	0	1	0	0	-1	0	1	-1	-1	-3	0	-2	-3	1	3				
T	1	-1	0	0	-2	-1	0	0	-1	0	-2	0	-1	-2	0	1	3			
W	-6	2	-4	-7	-8	-5	-7	-7	-3	-5	-2	-3	-4	0	-6	-2	-5	17		
Y	-3	-4	-2	-4	0	-4	-4	-5	0	-1	-1	-4	-2	7	-5	-3	-3	0	10	
V	0	-2	-2	-2	-2	-2	-2	-1	-2	4	2	-2	2	-1	-1	-1	0	-6	-2	4

Match 1 is the better match. The matching amino acids are the same between match 1 and match 2, but the 4 mismatches (italicized below) in both alignments are very different. Based on the Venn diagram, the mutations leading to the differences in match 1 seem more plausible. K to R substitutes one positive amino acid for another, while K to F substitutes a hydrophilic for a hydrophobic amino acid. The PAM250 matrix score also supports this assessment.

Match 1

Query: **F G Q V I P A K R**

Subjt: **F A N V M P A R E**

Match 2

Query: **F G Q V I P A K R**

Subjt: **F S C V F P A F V**

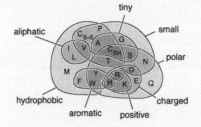

Match 1

Query: **F G Q V I P A K R**

Subjt: **F A N V M P A R E**

SCORE: 9+1+1+4+2+6+2+3−1=27

Match 2

Query: **F G Q V I P A K R**

Subjt: **F S C V F P A F V**

SCORE: 9+1−5+4+1+6+2−5−2=11

	A	R	N	D	C	Q	E	G	H	I	L	K	M	F	P	S	T	W	Y	V
A	2																			
R	-2	6																		
N	0	0	2																	
D	0	-1	2	4																
C	-2	-4	-4	-5	4															
Q	0	1	1	2	-5	4														
E	0	-1	1	3	-5	2	4													
G	1	-3	0	1	-3	-1	0	5												
H	-1	2	2	1	-3	3	1	-2	6											
I	-1	-2	-2	-2	-2	-2	-2	-3	-2	5										
L	-2	-3	-3	-4	-6	-2	-3	-4	-2	2	6									
K	-1	3	1	0	-5	1	0	-2	0	-2	-3	5								
M	-1	0	-2	-3	-5	-1	-2	-3	-2	2	4	0	6							
F	-4	-4	-4	-6	-4	-5	-5	-5	-2	1	2	-5	0	9						
P	1	0	-1	-1	-3	0	-1	-1	0	-2	-3	-1	-2	-5	6					
S	1	0	1	0	0	-1	0	1	-1	-1	-3	0	-2	-3	1	3				
T	1	-1	0	0	-2	-1	0	0	-1	0	-2	0	-1	-3	0	1	3			
W	-6	2	-4	-7	-8	-5	-7	-7	-3	-5	-2	-3	-4	0	-6	-2	-5	17		
Y	-3	-4	-2	-4	0	-4	-4	-5	0	-1	-1	-4	-2	7	-5	-3	-3	0	10	
V	0	-2	-2	-2	-2	-2	-2	-1	-2	4	2	-2	2	-1	-1	-1	0	-6	-2	4

Sweet Lou

To calculate the likelihood of amino acid substitutions, Dayhoff, the inventor of the PAM matrix, came up with a clever idea. Instead of deriving a complex formula that included biochemical properties and the genetic code, why not simply count how many times each amino acid mutated to every other amino acid using available protein sequence alignments? Dayhoff also calculated the number of times that each amino acid did not change at all.

To understand how counting can be used to calculate the future likelihood of certain events, consider an analogy from the sport of baseball. In baseball, one of the most difficult things to do is to hit the ball with the bat.[1] The best hitters manage to hit the ball effectively only 3 out of every 10 attempts and are even less likely to hit the best of all outcomes, a home run. Figure 7.1.1 shows the hitting statistics of the Detroit Tigers legend Lou Whitaker (nicknamed Sweet Lou). One way to determine how likely it would be for Sweet Lou to hit a home run would be to count how many times he actually hit a home run and compare that number to the number of times he did something else, including striking out, getting to first base, and walking. Based on his career stats, Sweet Lou hit a home run 3 out of every 100 times he tried to hit the ball.

To extend the baseball analogy a bit further, let's calculate a log odds score for Sweet Lou hitting a home run. We have the observed likelihood (0.03), but what is the expected likelihood? In this case, the expected likelihood might be the likelihood of a typical professional baseball player hitting a home run. If the typical player hits a home run 3 times out of every 1,000 times trying, then the expected likelihood is 0.003. If we calculate the log odds score using the base 10 logarithm,[2] then Lou Whitaker's log odds score of hitting a home run (S_{HR}) would be

$$S_{HR} = \log_{10}\left(\frac{0.03}{0.003}\right) = 1.0$$

meaning that Lou Whitaker is 10 times more likely to hit a home run than the typical professional player. If the value were −1.0, he would be 10 times less likely, and 0 means he would hit home runs at the typical rate [$\log_{10}(1) = 0$].

Similar to the baseball analogy, we could use protein sequence data and counting to determine the likelihood of a leucine (L) mutating to a tryptophan (W) or to a histidine (H) or not mutating at all. Dayhoff realized that she could look for mutations using protein multiple-sequence alignments. Both the Dayhoff PAM method and the BLOSUM method use counting to determine observed likelihoods, though the counting is done in a different manner for each method. In addition to the observed likelihoods, both methods also calculate an expected likelihood, which

Lou Whitaker
Position: Second Baseman
Bats: Left **Throws:** Right
5-11, 160lb (180cm, 72kg)
Born: May 12, 1957 in Brooklyn, NY

Summary	WAR	AB	H	HR	R	RBI	SB
Career	74.9	8570	2369	244	1386	1084	143

AB	R	H	2B	3B	HR	RBI	SB	CS	BB	SO
32	5	8	1	0	0	2	2	2	4	6
484	71	138	12	7	3	58	7	7	61	65
423	75	121	14	8	3	42	20	10	78	66
477	68	111	19	1	1	45	8	4	73	79

FIGURE 7.1.1. Using existing data to determine likelihoods: a baseball analogy. The figure shows 4 seasons of hitting statistics for Lou Whitaker (Sweet Lou), one of the finest second basemen ever to play the game. Using the data at the bottom left, we could calculate the likelihood that Lou would hit a home run (the HR column) the next time he came up to bat. To do this, we could count the total number of home runs (HR) Lou hit and divide that number by the total number of times he did anything else at bat, including striking out (SO), walking (BB), or getting a regular hit. In his career, Sweet Lou hit 244 home runs at 8,570 times at bat (AB), for a likelihood of ~0.03 (hit a home run 3 times out of 100 tries). The same type of calculation could be done to see how likely he was to get to first base, walk, or strike out. Photo courtesy of Aaron Caldwell, under license CC BY-2.0.

is the likelihood that this would have happened by random chance just based on the frequencies of the amino acids. The final score that is calculated for both methods is the log odds score, which is the log of the ratio of the observed likelihood to the expected likelihood. A positive log odds score indicates that the mutation is more likely than chance, while a negative score indicates that it is less likely than chance.

Calculating a PAM Matrix

The first step in calculating either a PAM or BLOSUM substitution matrix is to count the number of observed amino acid substitutions and the number of times amino acids did not change. To determine what substitutions occurred, Dayhoff constructed phylogenetic trees using a maximum-parsimony approach with the protein multiple-sequence alignments that existed at the time.

The first part of the approach is to calculate the mutation probability $M_{i,j}$, which is the likelihood that amino acid i will mutate into amino acid j. Figure 7.1.2 illustrates how a phylogenetic tree can be used to count amino acid mutation patterns for an alanine (A) to a cysteine (C), which can then be used to calculate $M_{A,C}$.

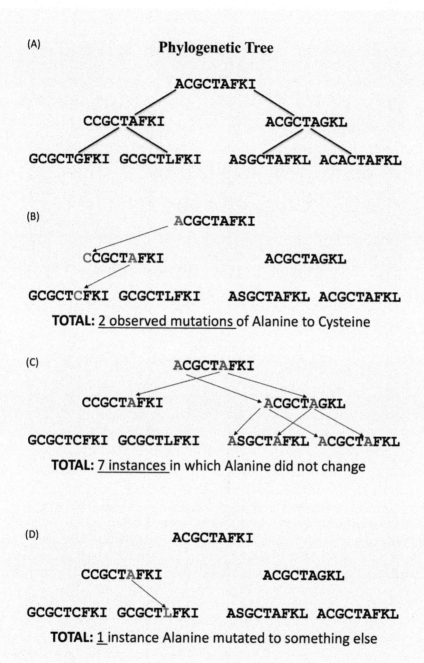

(A) **Phylogenetic Tree**

ACGCTAFKI

CCGCTAFKI   ACGCTAGKL

GCGCTGFKI GCGCTLFKI   ASGCTAFKL ACACTAFKL

(B)

ACGCTAFKI

CCGCTAFKI   ACGCTAGKL

GCGCTCFKI GCGCTLFKI   ASGCTAFKL ACGCTAFKL

TOTAL: 2 observed mutations of Alanine to Cysteine

(C)

ACGCTAFKI

CCGCTAFKI   ACGCTAGKL

GCGCTCFKI GCGCTLFKI   ASGCTAFKL ACGCTAFKL

TOTAL: 7 instances in which Alanine did not change

(D)

ACGCTAFKI

CCGCTAFKI   ACGCTAGKL

GCGCTCFKI GCGCTLFKI   ASGCTAFKL ACGCTAFKL

TOTAL: 1 instance Alanine mutated to something else

FIGURE 7.1.2 **Counting alanine substitutions using a phylogenetic tree. (A)** Phylogenetic tree reconstruction of the relationship of 4 aligned sequences. The tree has been inverted so that the root of the tree is at the top, with the 4 sequences from the alignment at the bottom. The tree includes 3 ancestral reconstructed sequence reconstructions, one at the root (top) and two descendants (middle), giving rise to the 4 sequences (bottom). **(B)** Counting the number of times in this data that alanine mutated to cysteine. **(C)** Counting the number of times alanine remained unchanged. **(D)** Counting the number of times alanine changed to another amino acid (in this case, leucine).

	A	R	N	D	C
A	9867	2	9	10	3
R	1	9913	1	0	1
N	4	1	9822	36	0
D	6	0	42	9859	0
C	1	1	0	0	9973

TOTAL = 10,000 events

A to A : 9867 events

A to C & C to A : 4 events

$M_{A,A} = 0.9867$ $M_{A,C} = 0.0004$

FIGURE 7.1.3. Observed amino acid substitution patterns for 10,000 events. The red boxes indicate the mutations of A to C and C to A, which were combined to determine the mutation rate. The PAM matrix also calculated the rate of not mutating. A did not change 9,867 times, making $M_{A,A}$ = 9,867/10,000 = 0.9867.

Figure 7.1.3 shows a part of the original substitution table used by Dayhoff to create the PAM scoring matrix. The data set available in the late 1970s was very limited because few sequences were available at the time for generating the observed and expected probabilities using the phylogenetic counting approach like the one shown in Fig. 7.1.2. To make the calculations more meaningful and easier to calculate, Dayhoff scaled the substitutions such that each amino acid underwent 10,000 total events. The matrix was also made symmetrical, such that $M_{i,j} = M_{j,i}$ by combining the counts of *i* to *j* and *j* to *i* and then dividing this over the total number of events (in this case, 10,000). For example, in Fig. 7.1.3, there is 1 observed A-to-C mutation and 3 C-to-A mutations, for a total of 4. Thus, $M_{A,C} = M_{C,A}$ = 4/10,000 = 0.0004.

Finally, to calculate the log odds score, one takes the natural log of the observed mutation rate divided by the expected rate as

$$S_{i,j} = \log \frac{p_i \times M_{i,j}}{p_i \times p_j} = \log \frac{M_{i,j}}{p_j} = \log \frac{\text{observed frequency}}{\text{expected frequency}}$$

where $S_{i,j}$ is the log odds score for a substitution of amino acid *i* to *j*, p_i and p_j are the frequencies of amino acid *i* to *j* respectively, and $M_{i,j}$ is the mutation rate. These log odds scores were then calculated for the transition of every amino acid to every other amino acid.

The original PAM matrix was based on a data set that includes 71 families of closely related proteins. In order to account for substitution patterns among more distantly related proteins, Dayhoff also introduced a scaling factor that projected the mutation rates for more distantly related sequences. Different matrices were then created for higher levels of protein divergence. A PAM1 matrix assumed an average of 1 amino acid substitution per 100 amino acids, while a PAM250 matrix assumed an average of 250 substitutions per amino acid (each amino acid mutated multiple times).

Calculating a BLOSUM Matrix

As previously mentioned, the log odds scores of the BLOSUM matrix are highly similar in principle to those of the PAM matrix. The scores are also based on the log ratio of the observed mutation rate to the expected mutation rate. However, instead of building a phylogenetic tree, the creators of the BLOSUM matrix, Henikoff and Henikoff, used the sequence alignments directly. Figure 7.1.4 shows the counting step of the algorithm. Using alignment columns in conserved blocks of multiple-sequence alignments, the first stage is to count all the possible amino acid pairings, referred to as tuples.

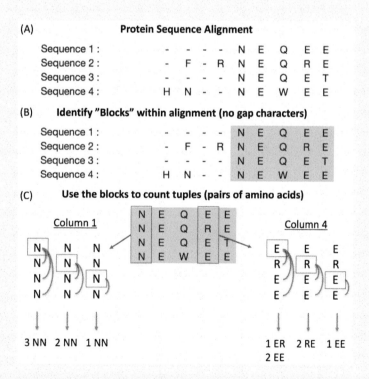

FIGURE 7.1.4. Counting amino acid tuples (pairs) in sequence alignment blocks.
(A) Multiple-sequence alignment of four proteins. **(B)** The first step of the BLOSUM algorithm is to identify blocks of positions in the sequence alignment without gaps. **(C)** Using these blocks, the algorithm counts all the possible pairs within each column. These are the observed substitutions. In the two columns analyzed in the figure, the algorithm counted 6 NN tuples in the first column and 4 EE tuples, 1 RE tuple, and 1 ER tuple in the second column.

These tuples are then used to calculate the observed frequencies of the pairings (Fig. 7.1.5). Then the frequency of the individual amino acids in the tuples is used to calculate the expected frequency of all the tuples (Fig. 7.1.6). Finally, with the observed and expected frequencies calculated, the last step is to calculate the log odds scores as follows:

$$\text{log odds ratio} = 2 \times \log_2\left(\frac{P(O)}{P(E)}\right)$$

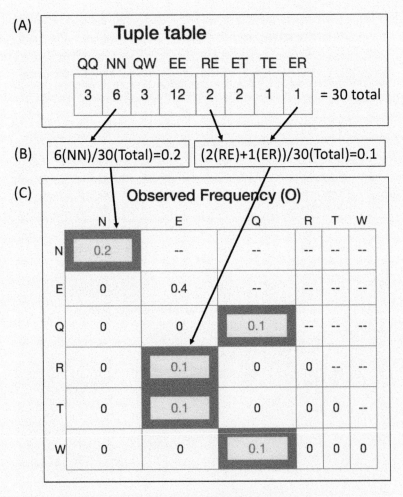

FIGURE 7.1.5. **Calculation of observed frequencies [P(O)]. (A)** The tuple table with the total number of pairs. **(B)** Calculation of observed frequencies. If the amino acid tuple count is of the same amino acid (i.e., NN), the tuple count is divided by the total number of observed tuples. If the amino acid count is of two different amino acids (i.e., RE), one first combines the total of both possible combinations (RE and ER) and then divides by the total number of observed tuples. **(C)** Table of observed frequencies. The zeros indicate tuples that have not yet been observed.

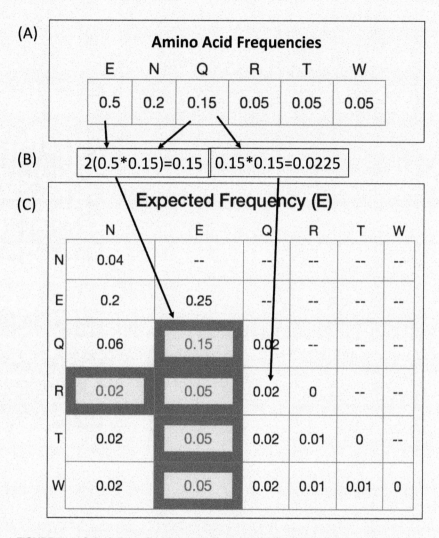

FIGURE 7.1.6. Calculation of expected frequencies [P(E)]. (A) Frequencies of each amino acid in the conserved sequence blocks. Glutamic acid (E) comprises half (10/20 = 0.5) of the total amino acids in the tuples from the highlighted sequence block in Fig. 7.1.3. **(B and C)** The expected values of the tuples are calculated by multiplying the independent probabilities (frequencies) of each amino acid. For instance, the expected frequency of a QQ pair is the square of their independent frequencies. Since the matrix is symmetrical, the expected frequency of an EQ pair is the combined frequency of EQ and QE.

BLOSUM matrices, like the PAM matrices, have also been designed for sequences with various levels of divergence. Unlike the PAM matrices, larger BLOSUM numbers should be used with more similar sequences. The BLOSUM80 matrix was constructed using protein sequence alignments that clustered together at the 80% similarity level, while the BLOSUM62 matrices clustered at 62%.

Exercises

Interactive exercise (theory)
Use the online PAM and BLOSUM interactive links below to learn how these matrices are created using observed amino acid substitutions. The interactive links explain how to use the teaching interactives. Once you learn how they work, use them to solve the problems in the next section.

BLOSUM Interactive Link
Link:
**http://kelleybioinfo.org/algorithms
/default.php?o=13**

PAM Interactive Link
Link:
**http://kelleybioinfo.org/algorithms
/default.php?o=14**

Problems

Solving for a BLOSUM matrix

1. Circle the sequence block in the multiple-sequence alignment below that could be used for BLOSUM matrix calculations.

Sequence 1 :	-	P	-	M	Q	K	Y	K	F	-	F	W	G
Sequence 2 :	E	A	V	-	Q	K	Y	K	N	-	-	-	A
Sequence 3 :	-	Q	K	-	Q	K	Y	K	W	F	-	M	V
Sequence 4 :	G	R	S	-	Y	K	Y	K	F	-	-	-	-

2. Fill in the blanks using the tuple values in the table.

Tuple table

Amino Acid Pair	QQ	WF	FW	QY	NF	YY	KK	FF	FN	NW
Count	3	1	1	3	1	6	12	1	1	1

Observed Frequency (O)

	Q	K	Y	F	N	W
Q	Fill in	--	--	--	--	--
K	0	0.4	--	--	--	--
Y	Fill in	0	0.2	--	--	--
F	0	0	0	Fill in	--	--
N	0	0	0	Fill in	0	--
W	0	0	0	Fill in	0.03	0

3. Fill in the blanks using the expected-probability values in the table.

Expected Probability (P)

Amino Acid	F	K	N	Q	W	Y
Expected Probability (P)	0.09	0.4	0.045	0.15	0.045	0.25

Expected Frequency (E)

	Q	K	Y	F	N	W
Q	0.02	--	--	--	--	--
K	0.12	0.16	--	--	--	--
Y	Fill in	0.2	0.06	--	--	--
F	0.03	0.07	0.05	Fill in	--	--
N	Fill in	0.04	Fill in	Fill in	0	--
W	0.01	0.04	0.02	0.01	0	0

Lab Exercises (Practice)

In this part of the exercise, you will explore how PAM and BLOSUM are used in BLAST protein searches. You will also learn how to use the hidden Markov model-based Pfam database.

Blastp Tutorial Link
Link:
http://kelleybioinfo.org/algorithms /tutorial/TProb1.pdf

Sample and lab exercise data:
http://kelleybioinfo.org/algorithms /data/DProb1.txt

Pfam Tutorial Link
Link:
http://kelleybioinfo.org/algorithms /tutorial/TProb2.pdf

Sample and lab exercise data:
http://kelleybioinfo.org/algorithms/data /DProb1.txt

Lab Exercise

Part 1. BLOSUM and PAM: using blastp advanced parameters

1. Use the protein BLAST (blastp) program at NCBI to perform a BLAST search with different PAM and BLOSUM matrices.

   a. Search using the prot1 sequence from the sample data. Adjust the blastp advanced parameters so that the search is performed with the PAM250 matrix. Find a result with an identity of less than 85%. Write the answers to the problems below.

   Protein function:

   NCBI reference sequence identifier:

   Organism (scientific name, and common name if available):

   Identities:

   Positives:

   Gaps:

   First 5 positions of the alignment of Query to Sbjct:

   b. Repeat the search with the prot1 sequence, but this time perform the search with the BLOSUM90 matrix. Find the same match result as in part a and write the answers to the problems below.

   Identities:

   Positives:

   Gaps:

   First 5 positions of the alignment of Query to Sbjct:

   c. Note any differences between the two BLAST alignments.

Part 2. Pfam database

Search the Pfam database with the prot2 sequence from the sample data and write/draw the answers to the following questions.

1. What significant Pfam match or matches did you find? Write the description.

2. What are any known functions of the protein family (families)?

3. What are the expected value (E values) of this match from the sequence search results page?

4. Draw/show the first two positions of the HMM logo.

5. What does the logo say about potentially important amino acids in this protein?

Notes

1. Not surprising since a hard ball about the size of a human fist is hurled at the batter at speeds close to 100 miles per hour (160 km/h), often with a wicked spin.
2. The PAM matrix calculates the log odds score using the natural logarithm, while BLOSUM uses the base 2 logarithm. It turns out that the base used is not all that important, but I find base 10 easiest to interpret.

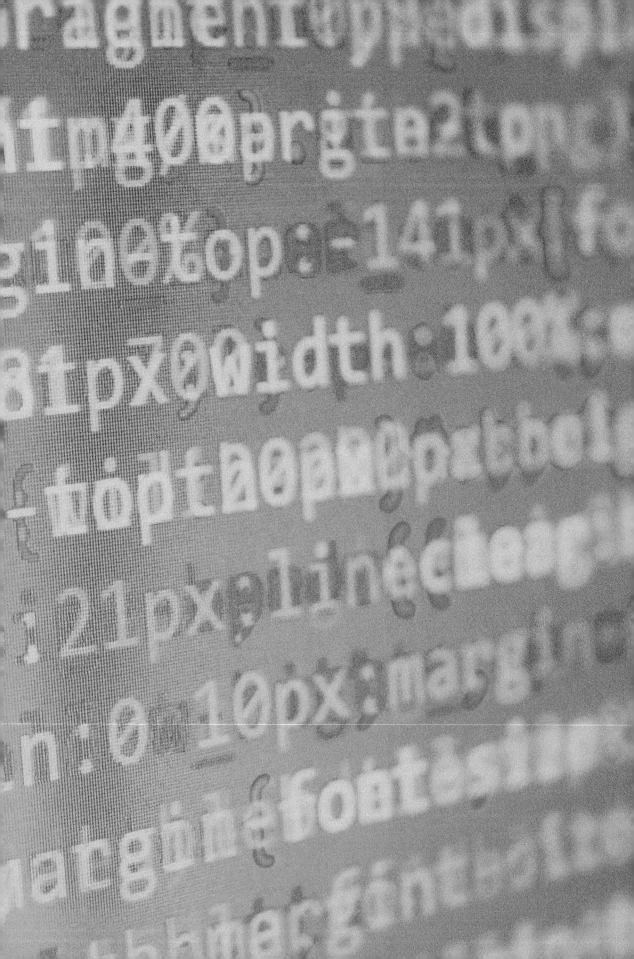

CHAPTER 08

BIOINFORMATICS PROGRAMMING: A PRIMER

The main goals of this hypertextbook are to teach the purpose of bioinformatics, the algorithms underlying some of the more commonly used bioinformatics programs, and how to use bioinformatics software to analyze sequence data. The final chapter focuses on the next step in the evolution of the bioinformatician: programming. While there are many terrific programs already available for bioinformatics analysis, there comes a time when you may need to reformat a large data set for a particular program, or you may need to compile a Unix program, or you may decide that cutting and pasting data sets into websites will delay your graduation or publication by approximately a decade. The purpose of this chapter is to familiarize you with a widely used bioinformatics programming environment, namely, the Unix operating system (OS), and two commonly used bioinformatics programming languages: the R and Python languages. Many excellent books, online tutorials, and classes exist for teaching Unix, R, and Python, and after finishing the exercises in this chapter you should be ready to tackle some of these resources and learning on your own. In the primers below you will find links to free resources and useful textbooks for additional self-tutorials or instructions.

The Unix Operating System

The collection of software known as the OS runs all aspects of a computational device (computer, phone, tablet, etc.). The OS controls and directs all the processors and devices, it runs the application software, and it contains and controls the computer file structure. The most commonly used computer OSs, namely, Windows and the MacOS, have very friendly user interfaces that make it easy to run programs, store data, and operate external devices. Smartphones are even easier. While these interfaces make standard tasks like opening a spreadsheet and storing photos simple, the windows-style interfaces make for poor programming environments.

Unix-style OSs are not user friendly, but they are dynamite programming environments. In a Unix system, one types specific instructions on the command line into a special window called a terminal to make things happen. For instance, instead of using the menu bar to make a new folder (in Unix systems, folders are called directories) on your computer, you can quickly make one with a command called "mkdir" (make directory). Let's say you had a lot of secrets to keep: you could make a directory called "TooManySecrets."

$ mkdir TooManySecrets

Then, instead of clicking or touching the TooManySecrets directory on the computer to open it and see the contents inside, you could instead use the cd command, which stands for "change directory," followed by the ls command, which lists the files.

$ cd TooManySecrets
$ ls

It may seem awkward, but once you learn the commands and understand the OS, with one command you can locate any file or any program on the computer no matter what directory you are in. For example, instead of clicking through four different folders to find the files in TooManySecrets, you might do the following instead:

$ cd ~/Documents/ScottStuff/MISC/TooManySecrets

Ta-da! The / symbol tells the computer to look in a subdirectory. The ~ is the home directory. So, this command changes directory by following a path from the home directory to Documents to ScottStuff to MISC to TooManySecrets. To find the path to any directory, just type the command pwd (print working directory) and then press the Enter key. In a Windows or Mac-style OS, to get to the file TooManySecrets, you would need to open Documents, then open ScottStuff, then open MISC, and then search for the file in the directory. The cd command does the same thing without all the clicking.

Even better, the Unix command line allows you to execute any software program from anywhere on the computer, as long as that program is installed in the correct directory. For example,

$ python

This command finds and opens the Python program. If configured correctly, you can also open browsers, spreadsheets, or any other program. These are major advantages for bioinformatics programming, and every bioinformatics programmer needs to be competent with the Unix language.

A short Unix tutorial

This tutorial presents a short introduction to the Unix/Linux OSs. Unix is an OS developed in the early 1970s that is widely used by programmers. Linux[1] is an open-source version of the Unix OS developed in the 1990s. Many computers run this OS, and a lot of programming code has been developed to run on Unix-like systems, so it is good for bioinformaticians to be familiar with this OS. After the tutorial in this chapter, you can complete a more extensive tutorial at **http://www.ee.surrey.ac.uk/Teaching/Unix/**

You can also learn from this excellent, free online book, *The Linux Command Line*, by William Shotts:

http://linuxcommand.org/tlcl.php

There are also many other available online tutorials and books. The MacOS comes with a Unix terminal, which can be used to complete the tutorial exercise. If you are using Windows 10, a Bash shell command line tool is included for developers, but you may need to activate it.[2] Alternatively, you will need to download and install a Unix emulator or a virtual machine on Windows, or install a flavor of Linux on your personal computer.

Step 1: Open a terminal window

Linux/Unix Tutorial: The terminal window

Unix-like systems can be accessed by opening a terminal window and typing on the command line. MacOS is built using a "flavor" of Linux.

Unix-like systems have a clear hierarchical directory structure. Every file and folder (directory) in these systems is accessible from the command line, as long as you know the file "path" to what you want to find.

To start playing around in Unix, open a terminal window that should look something like the one below.

```
Last login: Mon Jul 24 08:45:01 on ttys000
~ $ █
```

Step 2: Entering commands

Linux/Unix Tutorial: Entering commands

In the terminal window, you type in commands and hit "Enter" and things happen! Here, I have typed the commands *pwd* and then *ls*.

The *pwd* command stands for "print working directory" and reports the directory path of the current directory. The *ls* command lists all the files and directories within the current directory.

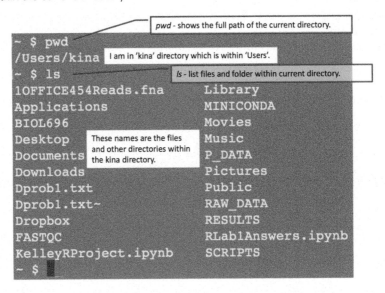

Step 3: More commands

Linux/Unix Tutorial: More commands

There are lots of commands you can use to navigate the operating system. Approximately 20 to 30 commands are used all the time. Unix is tedious at first, but it is much faster than clicking through a lot of folders and scrolling through windows. Below are some example commands.

Step 4: And even more commands

Linux/Unix Tutorial: And more commands

```
~ $ cd UNIX_EXAMP/
~/UNIX_EXAMP $ ls
my_data.txt       primer2.csv
~/UNIX_EXAMP $ head my_data.txt
>FWIRNKE01DKIF6 rank=0000177 x=1346.0 y=2772.0 length=53
CGATATTCGATCCGCATCGCTGCCCTACCCGTGGAGTGCCTCCCTCGGNGCAG
>FWIRNKE01CDBE3 rank=0000320 x=854.0 y=2685.0 length=53
GCGAGCAGCAATCATGCTGCCTCCCGTCGGAGGTGGCCCTCCCCTCCCTCCGC
>FWIRNKE01BKZJJ rank=0000535 x=531.0 y=3933.0 length=45
CGAGCAGCACATCATGCCTGGCCTTCCGACGGAGTGCCTCCTCGC
>FWIRNKE01CT8MK rank=0000656 x=1047.0 y=1690.0 length=45
CGTATGACTGTATCATGCTGCCTCCCGTAGGAGTGCCTCCTCGAC
>FWIRNKE01EP6FI rank=0000658 x=1821.0 y=1148.0 length=42
CGAGCAGCACATCATGCTGCCTCCCAGGAGTGCCTCCCTCGC
~/UNIX_EXAMP $ mkdir SCOTT_RULES
~/UNIX_EXAMP $ ls -al
total 80
drwxr-xr-x    5 kina    staff     170 Jul 24 10:17 .
drwxr-xr-x+ 63 kina    staff    2142 Jul 24 10:10 ..
drwxr-xr-x    2 kina    staff      68 Jul 24 10:17 SCOTT_RULES
-rw-r--r--    1 kina    staff     526 Jul 24 10:08 my_data.txt
-rw-r--r--@   1 kina    staff   33682 Jul 24 10:09 primer2.csv
~/UNIX_EXAMP $ cp primer2.csv primer_copy.csv
~/UNIX_EXAMP $ ls
SCOTT_RULES      my_data.txt      primer2.csv      primer_copy.csv
~/UNIX_EXAMP $ rm primer_copy.csv
~/UNIX_EXAMP $ less primer2.csv
~/UNIX_EXAMP $ mv my_data.txt SCOTT_RULES/
~/UNIX_EXAMP $ ls
SCOTT_RULES      primer2.csv
~/UNIX_EXAMP $ cd SCOTT_RULES/
~/UNIX_EXAMP/SCOTT_RULES $ ls
my_data.txt
~/UNIX_EXAMP/SCOTT_RULES $ █
```

head - shows the first 10 lines of a file.

mkdir - makes a new (empty) directory.

cp - make a copy of a file. This command copies the contents of primer2.csv to primer_copy.csv

rm - stands for "remove" and it deletes files.

The *less* command lets you scroll through the contents of a file one page at a time.

mv – moves a file from one place to another. This command moves my_data.txt into SCOTT_RULES

Introduction to R

R is a programming language and environment for statistical computing and graphics. R has the tools of a commercial statistical package (e.g., SPSS, Systat, or SAS) but is free of charge (I wish I had discovered R years before I did.) I regularly use R to test for statistical correlations, make box plots, and perform analyses of variance, regressions, or dozens of other statistical tests. However, the reason that R is a must for bioinformatics is because R comes with many other packages (libraries), including many bioinformatics packages. Dozens of bioinformatics methods have been coded in R, and an R library is pretty much the first accessible place for a new sequence analysis algorithm. At the time of this writing, the open-source Bioconductor depository (**http://www.bioconductor.org**) had more than 1,200 R packages for high-throughput data analysis. The CRAN[3] depository also contains dozens of libraries[4] with powerful statistics and algorithms for biological data analysis, such as "vegan," "randomForest," "ggplot," and many multivariate analysis packages.[5]

This tutorial teaches a few basics to get you going, including (i) how to install R, (ii) how to upload a data file, and (iii) how to perform a few simple statistical analyses.

To continue learning R, I recommend the tutorial at

http://www.cyclismo.org/tutorial/R/index.html

Another more challenging tutorial can be found at

http://tryr.codeschool.com/levels/1/challenges/1

Finally, I recommend the excellent *R Cookbook*, by Paul Teetor.

Step 1: Installation instructions

R Tutorial: Installation

To install R, go to the R website: http://www.r-project.org

The R Project for Statistical Computing

[Home]

Download

CRAN

R Project

About R
Logo
Contributors
What's New?
Reporting
Bugs
Development
Site
Conferences
Search

R Foundation

Foundation
Board
Members
Donors
Donate

Getting Started

R is a free software environment for statistical computing and graphics. It compiles and runs on a wide variety of UNIX platforms, Windows and MacOS. To download R, please choose your preferred CRAN mirror.

If you have questions about R like how to download and install the software, or what the license terms are, please read our answers to frequently asked questions before you send an email.

News

> Click the download link.

- The R Journal Volume 9/1 is available.
- R version 3.4.1 (Single Candle) has been released on Friday 2017-06-30.
- R version 3.3.3 (Another Canoe) has been released on Monday 2017-03-06.
- The R Journal Volume 8/2 is available.
- **useR! 2017** (July 4 - 7 in Brussels) has opened registration and more at http://user2017.brussels/
- Tomas Kalibera has joined the R core team.

R Tutorial: Installation

Choose the closest CRAN Mirror site for download.

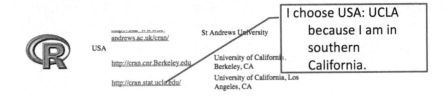

> I choose USA: UCLA because I am in southern California.

USA

andrews.ac.uk/cran/   St Andrews University

http://cran.cnr.Berkeley.edu   University of California, Berkeley, CA

http://cran.stat.ucla.edu/   University of California, Los Angeles, CA

The Comprehensive R Archive Network

Download and Install R

Precompiled binary distributions of the base system and contributed packages, **Windows and Mac** users most likely want one of these versions of R:

CRAN

Mirrors
What's new?
Task Views
Search

- Download R for Linux
- Download R for (Mac) OS X
- Download R for Windows

> And I have a Mac so this is my link

Step 2: Commands in R

R Tutorial: Opening the R console

Double click the R icon and you should get a window that looks like this:

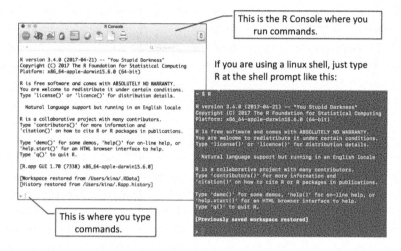

This is the R Console where you run commands.

If you are using a linux shell, just type R at the shell prompt like this:

This is where you type commands.

Step 3: Reading in a data set for statistical analysis

This next section requires that you download a data set and put it on your computer desktop.[6] The file used in the exercise can be found at

http://kelleybioinfo.org/algorithms/basics/programming/RTestData.txt

Notes

1. The sample names cannot start with a number. For instance, you cannot put 001 instead of S001. If you have numbers, put a letter in front of them.
2. Do not allow spaces in any of the names of variables. No funny symbols or special characters. Letters and numbers only.
3. Empty cells in a data set, called a data frame in R, must be replaced by NA (for "not available").

The next part loads the data into R. This is called the dataframe. In this case, read the data in using the read.table function and assign it to the variable d (for "data"). Note that many textbooks and tutorials assign values to variables using the arrow symbol. For instance, the code shown in step 2 could be written

d<-read.table("Desktop/RTestData.txt",header=TRUE)

However, most programming languages use the equal sign to assign variables. I think it looks much better than the arrow, and it works great in R. You can use arrow symbols if you like, but it takes twice as many keystrokes.

Step 4: Analyzing your data

After you have installed R and loaded up the test data set, try out some of the very exciting statistical analyses!

R Tutorial: Reading in a data set

Step 1: You need a text-only file that R can read. I created a tab-delimited text file called **RTest-Data.txt**. This data is from a gum disease study and has abundances of bacteria found in the human mouth.

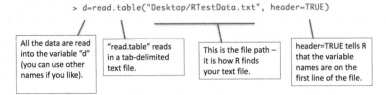

id	strep	prev	lepto	fuso	veil	time	status	pocket	deepest
S001	57.4	10.2	0.1	0	6.9	1	2	2.7	3.8
S001	26	0	25.6	0	6.3	2	2	2.7	3.0
S006	19	24.2	4.8	5.7	4.2	1	1	2.5	3.2
S006	15.2	4.2	0.2	2.6	3.4	2	1	2.4	3.2
S007	33.2	2.3	7.4	1.3	12	1	2	2.4	2.7
S007	18	0.3	13.6	1.7	13.8	2	2	2.4	2.8
S008	12.3	11.5	9.1	5.5	5.7	1	1	2.6	3.5
S008	3.3	24.7	6.3	10.4	3	2	1	2.5	3.2
S009	9.1	32.7	1.9	14.3	1.6	1	1	2.9	4.0
S009	22	8.9	17.8	4.8	14	2	1	2.5	3.5

Description of RTestData: Below is the meaning of each column header.

id=Code for each patient. Two rows for each subject: one before and one after gum cleaning.

strep=Percentage of *Streptococcus* bacteria

lepto=Percentage of *Leptotrichia* bacteria

prev=Percentage of *Prevotella* bacteria

fuso=Percentage of *Fusobacteria* bacteria

veil=Percentage of *Veillonella* bacteria

time=Time that sample was taken: 1–before gum cleaning; 2–after gum cleaning

status=Disease status: 1 is healthy, 2 is diseased gums

pocket=Average gum pocket depth across all the teeth in the mouth (in millimeters)

deepest=Depth of the deepest gum pocket in the mouth (in millimeters)

Example: 15.2% of Subject 6 (S006) are Streptococcus (strep) bacteria before treatment (time 1).

RTestData.txt can be downloaded by saving it as a text file or by copy/pasting data into a text file:
http://kelleybioinfo.org/algorithms/basics/programming/RTestData.txt

R Tutorial: Reading in a dataset (Mac and Linux)

Step 2: To read in your data set, you need to know where the data set is on your computer. (I made it easy and put it on my desktop.) Then type the path to the folder/directory in the console and hit return.

```
> d=read.table("Desktop/RTestData.txt", header=TRUE)
```

| All the data are read into the variable "d" (you can use other names if you like). | "read.table" reads in a tab-delimited text file. | This is the file path – it is how R finds your text file. | header=TRUE tells R that the variable names are on the first line of the file. |

Then, if you just type "d" and hit enter you will see your lovely data!

```
> d=read.table("Desktop/RTestData.txt", header=TRUE)
> d
    id strep prev lepto fuso veil time status pocket deepest
1 S001  57.4 10.2   0.1  0.0  6.9    1      2    2.7     3.8
2 S001  26.0  0.0  25.6  0.0  6.3    2      2    2.7     3.0
3 S006  19.0 24.2   4.8  5.7  4.2    1      1    2.5     3.2
4 S006  15.2  4.2   0.2  2.6  3.4    2      1    2.4     3.2
```

R Tutorial: Reading in a data set (Windows)

To read in your data set in Windows, you have to find the path to the file. To find the path, right-click the data file and choose "Properties" at the bottom of the menu. You will get a window that looks like this:

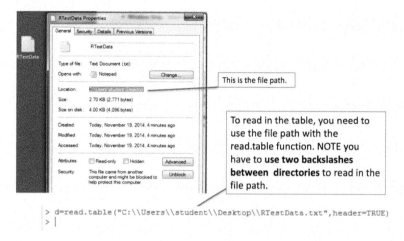

> d=read.table("C:\\Users\\student\\Desktop\\RTestData.txt",header=TRUE)
> |

R Tutorial: Simple analyses

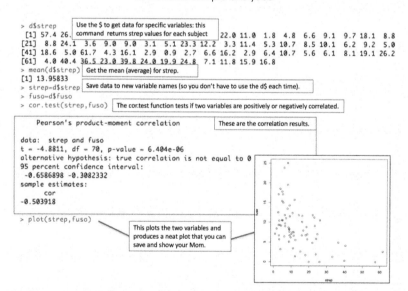

Note that after reading in the variable, to access the data associated with the variable, you must use the $. Because it is annoying to keep typing the dollar sign, I copied the data to a new variable name like this: **strep=d$strep**

Introduction to Python

Python has become the most widely used programming language in bioinformatics, and for good reason. Not only is Python a flexible, fully functional programming language used in applications around the world, but also it was designed to be relatively easy to learn. Guido van Rossum,[7] Python's creator, combined the style of the C programming language with the ease of a simpler learning language called ABC. The result was a clean and rapid scripting language with a simple syntax that results in fewer bugs. Python really shines in bioinformatics because it is wonderful for opening, reading, and parsing data files, and because it is a tremendous so-called glue language. With Python it is easy to glue together many different programs (R, Unix, C, and Python) into a powerful analysis workflow.

The Python exercises in this chapter assume the use of PYTHON 3.x version of the programming language. The current version as of this writing is Python 3.6. Python is available for download and installation at

https://www.python.org/

if it is not already on your system (check this). You can read more about why Python is awesome at

https://www.python.org/about/success/

The tutorial here includes a short primer on using the *Python Interpreter* and an example of how to write, save, and execute a simple programming file. If you want to learn more after finishing this, there are many excellent free tutorials, including on the Python website.

Python tutorial:
https://docs.python.org/3/tutorial/index.html
Python for total beginners:
https://wiki.python.org/moin/BeginnersGuide/NonProgrammers
https://automatetheboringstuff.com/

There are loads of other resources and books for learning Python, including a book called *Python for Biologists*, by Martin Jones.

Step 1: Invoking Python

Python on the command line

After installing Python, you should be able to invoke Python on the command line in a terminal window by typing the name of the program. This opens the *Python Interpreter*, where you can run Python code directly in the terminal by typing 'python' and hitting the *Enter* key.

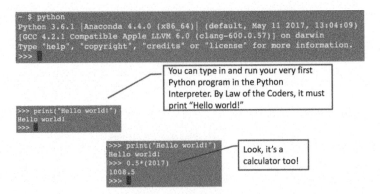

This part assumes that you have installed a version of Python 3 and can run the program in a terminal window. There are other ways to run Python, including with a text editor like Emacs.

Step 2: Running a loop

The Python Interpreter: Instant feedback

Let's do something a little more interesting. This code snippet prints the numbers from 5 to 9. How would you change the loop to print numbers from 3 to 101? Try it yourself!

> This statement is called a *for loop*, which loops through a set of conditions one at a time from the beginning to the end and does something. In this case, the *for loop* loops through a list of five numbers and prints each of them to the screen.

```
>>> print("Let's print some numbers in the terminal window!")
Let's print some numbers in the terminal window!
>>> for i in range(5,10):
...     print(i)
...
...
5
6
7
8
9
>>>
```

> Both *range* and *print* are python functions. The *range* function creates a list of numbers, while the *print* function dumps output to the screen.

> Notice that the *print* statement is within the *for loop*. Python knows it is within the loop because it is tab-indented.

Step 3: Saving a Python program to a file

This part shows you how to save a file and execute it using Python on the command line. Most people save their Python files by adding a ".py" file extension at the end of the name. This allows you to remember that it is indeed a Python program file, and many programs will also recognize this as a Python file by the file extension. You will need to use a text editor to save your file; there are many to choose from, including editors that come with your OS, such as Notepad or nano.[8]

Python: Saving your programming code

Once you quit the Python interpreter (type *Control-d* to quit) all your work disappears. The interpreter is nice for testing out simple functions or for doing calculations, but your work is lost after you quit. To save your program, you need to write it in a separate file. Below we save our work with the Emacs text editor, a commonly used text editor for programming. (The *nano* program is an editor that often comes with Linux. Type *nano* at the prompt.

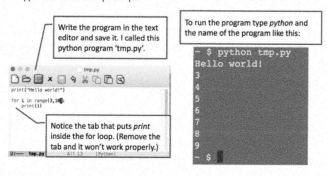

> Write the program in the text editor and save it. I called this python program 'tmp.py'.

> To run the program type *python* and the name of the program like this:

> Notice the tab that puts *print* inside the for loop. (Remove the tab and it won't work properly.)

After you finish these primers you can work your way through the excellent websites, online pdfs, and books mentioned throughout the chapter. Armed with this knowledge, you are on your way towards rapid, custom-designed high-throughput analysis of your own crazy data. Both R and Python have scores of freely available bioinformatics-specific libraries that will allow you to implement the algorithms found in the book and much more—including parsing data sets and analyzing new data.

Biopython: http://biopython.org
Bioconductor: https://www.bioconductor.org

Certainly, one of the best ways to truly learn a programming language is to have a project that you really want, or really need, to do. Perhaps you need to reformat a data set to run with specialized analysis software (Python), or you need to perform stepwise multiple regression analysis (R), or maybe you need to create a rapid workflow that uses a series of unrelated programs (Unix). Programming can be a challenging, error-filled journey, but it's worth it when you experience that feeling of glee as your code crunches through thousands of data points to produce an analysis of unsurpassed beauty. Then you're ready to take on the world!

Notes

1. Linux was created by a bloke named Linus Torvald, ergo "Linus's Unix," shortened to "Linux."
2. **http://www.windowscentral.com/how-install-bash-shell-command-line-windows-10**
3. From the R website: "CRAN is a network of ftp and web servers around the world that store identical, up-to-date, versions of code and documentation for R."
4. Inside R packages can be readily installed by using the install.package command. For example, to install ggplot2, use install.packages("ggplot2"). To use the new library, type library-(ggplot2).
5. **https://cran.r-project.org/web/views/Multivariate.html**
6. To save this file from a web browser, you can, for example, click "Save As" to save as a text file in Firefox, or click "Save As" and select "page source" in Safari. Also, you can copy/paste into a text-only processor like nano or Emacs. To check that it is a text file, you can use the command *head* or *less*. If you see a lot of garbage, it is not text-only. Note that Microsoft Word is not a text-only processor.
7. van Rossum was a massive Monty Python fan, hence the name: Python.
8. I prefer the Emacs editor that can be installed for all OSs, though it does have a steep learning curve.

INDEX

Adenine, nucleotide base in DNA and pairing, 9, 10
Alanine, propensities for, 59
Alignment page, 1–3
Amino acids
    baseball analogy for likelihood in substitution,
        165–166
    chemical structures of, 16
    free energies for transfer of, 53
    PAM and BLOSUM matrices, 160–161
    propensities, 59, 61
    substitution bias, 159–160
    substitution matrices, 157–159
    Venn diagram of properties and genetic code, 160
Androgen receptor, 136
Anthrax toxin, representation of, 48
Archaea, phylogenetic tree of life, 134
Asparagine, propensities for, 59

Bacteria, phylogenetic tree of life, 134
Bacterial gene, general structure of, 19
Baseball analogy, likelihood in amino acid substitution,
    165–166
Bioinformatics, 5
    computer and, 6–8
    hidden Markov Models (HMMs), 161–162
    methods, 48–50
    power of, 33–35
    protein, 47–48
Bioinformatics software, 1–3
    Python programming language, 187–190
    R programming language, 183–187
    Unix operating system, 179–183
Biological databases and data storage
    concepts, 21–25
    exercises, 25–28
    learning objectives, 20–21
    motivation, 20
Biological molecules, properties, 5

BLAST (Basic Local Alignment Search Tool) algorithm, 1
    algorithm activity, 36–38
    BLAST It, 31–32
    BLAST tutorial, 40
    characterizing protein in *Escherichia coli*
        genome, 35
    concepts of, 36–38
    interactive exercise, 38, 39
    lab exercises, 40–44
    learning objectives, 36
    massive parallelization of, 32, 34
    motivation, 36
    power of, 33–35
    results of search of hypothetical FX093345
        nucleotide, 32, 33
    sequence alignment, 158
Blastp Tutorial Link, 175, 176
BLOSUM (Blocks Substitution Matrix), 157–159
    activity generating PAM and, 163–166
    calculation of, 169–171
    PAM and, 160–161
BLOSUM Interactive Link, 172
Bootstrap
    analysis, 138
    phylogenetics, 137–139
    term, 139n7

Chernobyl chicken, sequence, 100–101
Chou-Fasman algorithm, 50, 58, 157
Chou-Fasman Interactive Link, 61
Clustal Omega alignment program, 72, 74, 85–86, 88
Clustal Omega Tutorial Link, 83
ClustalW program, 72
Computer. *See also* Bioinformatics software
    bioinformatics, 6–8
    DNA in the, 8–11
    protein sequences in the, 13–14, 16–17
    RNA in, 11–13

Cyanobacteria, 133
Cytosine, nucleotide base in DNA and pairing, 9, 10

Data storage, biological databases and, 20–28
Deoxyribonucleic acid (DNA)
    chemical structure at atomic level, 8
    computational translation of DNA sequence, 49
    in the computer, 8–11
    double helix structure of, 111, 112
    multiple sequence alignment of, 69
    progressive sequence alignment of four DNA
        sequences, 72
    reverse complementation of DNA sequence, 11
    ribonucleic acid (RNA) *vs.*, 111, 112
    sequence motifs, 92
    sequence showing four nucleotide bases and
        pairing, 9
Distance Matrix Interactive Link, 149
DNA. *See* Deoxyribonucleic acid (DNA)
DNA Learning Center (DNALC), 12
Double helix, deoxyribonucleic acid (DNA), 111, 112
Dynamic programming
    activity, 74–78
    concepts, 74–76
    lab exercises, 83
    learning objectives, 74
    motivation, 74
    solving alignment problem, 77–78

Escherichia coli, 133
    characterizing novel bacterial protein in
        genome, 35
    conserved sequences in promoter regions of, 93
    sequence alignment of regions in *E. coli*
        genome, 92
Estrogen receptor, 136
Eukaryota, phylogenetic tree of life, 134
Eukaryotes
    general structure of gene, 18
    genetic code for, 17
    simplified illustration of RNA transcription in, 12

FASTA protein sequence, 27, 28

GenBank, 7, 20, 21, 48
Gene, molecular structure of, 17–19
Genetic information, molecular structure of a gene,
    17–19
Glucocorticoid receptor, 136
Guanine, nucleotide base in DNA and pairing, 9, 10

Hesper, Ben, 72
Hidden Markov Models (HMMs)
    bioinformatics, 161–162
    definition, 162
Hogeweg, Paulien, 72
Human estrogen receptor, 50
Human genome, small fraction of, 20

Humans, DNA sequence alignment, 67, 68
Hydrophobicity Interactive Link, 54
Hydrophobicity plotting
    activity, 52–55
    concepts, 52–54
    lab exercises, 56–57
    learning objectives, 52
    motivation, 52

Jones, Martin, 188

Kelley Bioinformatics
    Alignment page, 1–3
    website, 1

Likelihood, amino acid substitution, 165–166
Linux, 180, 190n1
The Linux Command Line (Shotts), 181
Lysines, calculation of propensity, 59, 60

Markov models, 48, 56, 70, 93n2
    hidden Markov models (HMMs), 161–162
MatrixPlot, predicting RNA structure, 128, 130
MatrixPlot (Mutual Information) Tutorial Link, 128
Maximum Parsimony Interactive Link, 149
Methanobacterium, 133
Mfold (Free Energy) Tutorial Link, 128
Mineralocorticoid receptor, 136
Molecular biology, 5
Mouse, DNA sequence alignment, 67, 68
Multiple sequence alignments (MSAs), 67, 72–73
    DNA and protein sequences, 69
    lab exercises, 84–88
    progressive sequence alignment of four DNA
        sequences, 72
Mutations, retroviruses, 73n1
Mutual Information (MI)
    algorithm for predicting RNA structure, 115–117,
        121–124
    exercise, 127
    principle of, 124
Mutual Information Interactive Link, 125
Mycobacterium, 26
Mycobacterium tuberculosis, 136
Myosin, representation of, 48

National Center for Biotechnology Information (NCBI),
    20–21, 25
    BLAST tool, 40–44
    exercise for *Sulfolobus solfataricus*, 28
    exercise using NCBI PubMed, 26
    NCBI ORF finder, 43
Needleman-Wunsch DNA/Protein Alignment
    Interactive Link, 79
Needleman-Wunsch method
    dynamic programming algorithm, 74, 76, 80–81
    screenshot of module, 158
    substitution matrices, 157

Nucleosome core particle:DNA fragment complex, 48
Nucleotide bases, pairing in DNA sequence, 9, 10

Occam's razor, 154n4

Pace, Norman, 133
PAM (Point Accepted Mutation)
    activity generating BLOSUM and, 163–166
    amino acid substitution patterns, 168
    BLOSUM and, 160–161
    calculation of, 166–169
    substitution matrix, 157–159
PAM Interactive Link, 172
Parsimony, 154n4
Patterns in data, 91–92
    logos for sequences in critical cell functions, 93
    sequence alignment of regions in *Escherichia coli* genome, 92
    sequence motifs, 92–93
Pfam (protein family) database, 48
Pfam Tutorial Link, 175, 177
Phylogenetic analysis
    activity, 140–148
    concepts, 140–141
    distance method, 141–143
    interactive exercises, 149–151
    lab exercises, 152–153
    learning objectives, 140
    maximum parsimony (MP) method, 143–148
    motivation, 140
Phylogenetic Analysis Tutorial Link, 152
Phylogenetics, uses of, 135, 136
Phylogenetic tree, 133
    aspects of, 137
    bootstrap analysis, 138
    bootstrap method, 137–139
    counting alanine substitutions using, 166, 167
    interpretation of, 136–137
    maximum parsimony (MP) method, 145–148
    set theory, 146, 154n7
    universal, 134
    using distances to build a, 142–143
Phylogeny, 133–134, 136, 140
Position-specific weight matrices
    activity, 102–108
    concepts, 102–105
    lab exercises, 107–108
    learning objectives, 102
    motivation, 102
Probability
    hidden Markov Models (HMMs), 161–162
    protein substitution matrices, 157–159
    substitution bias, 159–160
Progesterone receptor, 136
Progressive alignment method, 72

Propensities
    calculating for amino acids, 59–61
    for lysines, 60
Protein
    bioinformatics, 47–48
    cellular processes, 48
    chemical structures of amino acids for, 16
    computational translation, 49
    hydrophobicity plotting activity, 52–55
    hydrophobic regions in, 50
    sequence motifs, 92–93
    structure aspects, 15
    translation, 13
Protein secondary structure
    calculating propensities, 59–61
    concepts, 58–59
    lab exercises, 63–65
    learning objectives, 58
    motivation, 58
    prediction activity, 58–62
Protein sequence
    in the computer, 13–14, 16–17
    FASTA, 27, 28
    multiple sequence alignment of, 69
Protein sequence motifs
    activity, 94–101
    concepts, 94–96
    lab exercises, 98–101
    learning objectives, 94
    motivation, 94
ProtParam tool, 64
ProtScale Tutorial Link, 56, 63
Python for Biologists (Jones), 188
Python programming language, 187–190
    tutorial, 188

Retroviruses, 73n1
Ribonucleic acid (RNA)
    calculating RNA free energy, 120
    compensatory mutations in RNA structure, 116
    in the computer, 11–13
    CRISPR-Cas9 system, 113
    deoxyribonucleic acid (DNA) *vs.*, 111, 112
    folding algorithms, 115–117
    folds for sequence, 115
    messenger RNA (mRNA), 13
    predicting structure, 114–117
    ribosomal subunits, 113
    roles of, in cells, 112–114
    secondary structure, 113
    secondary-structure diagram of RNase P structural RNA, 114
    sequence motifs, 92
    simplified illustration of transcription in eukaryotes, 12
    simplified illustration of translation, 14
    single-stranded nature of, 111
    spliceosome, 113
    transfer RNA, 112, 113

Ribonucleic acid (RNA) structure prediction
activity, 118–128
calculating RNA free energy, 120, 121
comparing possible structures, 121
concepts, 118–124
lab exercise, 129–130
learning objectives, 118
motivation, 118
mutual information (MI) method, 115–117, 121–124
thermodynamic secondary-structure prediction, 115, 118–121
RNA. *See* Ribonucleic acid (RNA)
RNA Free-Energy Interactive Link, 125
R programming language, 183–187
tutorial, 183–187

ScanProsite Tutorial Link, 98
Sequence alignment, 67
challenges in, 70
issues in, 70–71
mutational history of protein-coding gene, 71
nature's experimental results, 67–70
position-specific weight matrices (PSWMs), 102–108
progressive, of four DNA sequences, 72
Sequence Motif Interactive Link, 96
Set theory, 146, 154n7
Severe acute respiratory syndrome (SARS), 33, 136
Shotts, William, 181
SMART (Simple Modular Architecture Research Tool), 100
Smith-Waterman variant of algorithm, 89n4
Steroid hormones, 50
Substitution matrices
activity generating PAM and BLOSUM, 163–171
calculating BLOSUM matrix, 169–171
calculating PAM matrix, 166–169
concepts, 163–166
determining substitution bias, 159–160
interactive exercise, 172–174
lab exercises, 175–177
learning objectives, 163
motivation for, 163
PAM and BLOSUM, 160–161
protein (amino acid), 157–159
Sulfolobus solfataricus, exercise using, 28

Taxonomy, 137
Thermodynamic secondary-structure prediction, RNA structure, 115, 118–121
Threonine, propensities for, 59
Thymine, nucleotide base in DNA and pairing, 9, 10
Torvald, Linus, 190n1
Transcription, 12, 19n5
Transcription Factor Binding Site Tutorial Link, 107
Transcription factors (TFs), 102
Transcription factors binding sites (TFBSs), 102, 107
Translation, 14, 19n5
Tree of life, 133
expanding, 135
ramifications of, 133–134

UniProt, 48
Unix operating system, 179–183
Linux/Unix tutorial, 181–183
tutorial, 180–181

Watson-Crick base pairings, 111, 119
Weight Matrix Interactive Link, 105
Whitaker, Lou, 165, 166
Woese, Carl, 133, 135
www.kelleybioinfo.org, 1